BestMasters

Mit „**BestMasters**" zeichnet Springer die besten Masterarbeiten aus, die an renommierten Hochschulen in Deutschland, Österreich und der Schweiz entstanden sind. Die mit Höchstnote ausgezeichneten Arbeiten wurden durch Gutachter zur Veröffentlichung empfohlen und behandeln aktuelle Themen aus unterschiedlichen Fachgebieten der Naturwissenschaften, Psychologie, Technik und Wirtschaftswissenschaften. Die Reihe wendet sich an Praktiker und Wissenschaftler gleichermaßen und soll insbesondere auch Nachwuchswissenschaftlern Orientierung geben.

Springer awards **"BestMasters"** to the best master's theses which have been completed at renowned Universities in Germany, Austria, and Switzerland. The studies received highest marks and were recommended for publication by supervisors. They address current issues from various fields of research in natural sciences, psychology, technology, and economics. The series addresses practitioners as well as scientists and, in particular, offers guidance for early stage researchers.

More information about this series at http://www.springer.com/series/13198

Matthäus Jäger

Fuel Tank Sloshing Simulation Using the Finite Volume Method

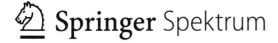 Springer Spektrum

Matthäus Jäger
Graz, Austria

ISSN 2625-3577   ISSN 2625-3615   (electronic)
BestMasters
ISBN 978-3-658-25227-4   ISBN 978-3-658-25228-1   (eBook)
https://doi.org/10.1007/978-3-658-25228-1

Library of Congress Control Number: 2019930096

Springer Spektrum

This Springer Spektrum imprint is published by the registered company Springer Fachmedien Wiesbaden
GmbH part of Springer Nature
The registered company address is: Abraham-Lincoln-Str. 46, 65189 Wiesbaden, Germany

Preface

A fuel tank system provides a wide range of applications for numerical simulations particularly in the area of fluid dynamics. There are many challenging topics and it took me some time to gain the necessary knowledge and develop the abilities for doing such simulations. I really had to learn that a successful simulation does not only require a good theoretical knowledge of the problem but also a good portion of patience and time. Therefore I am very grateful to all the people who supported me on this way.

First of all I want to thank you, Sandra, for your great encouragement and support this whole time. Special thanks go to my family and friends for enabling me my studies and making this time at university unforgettable.

Furthermore I want to thank Dr. Stefan Reiterer and my colleagues at Magna Steyr Fuel Systems for giving me the opportunity and the time to dive into such an interesting topic. Finally I want to thank my supervisor Professor Haase for his support and for giving me the freedom to do this thesis the way it is.

Thank you!

Contents

List of Figures

List of Tables

1 Introduction

The fuel tank system is a very important part of most vehicles, but it is
also one of the last things in the development process of an automobile.
The first fact results in high requirements with respect to stability,
durability and general performance of the tank. The latter, on the
other hand, brings limitations concerning the volume and geometry of
the tank, because the available design space for the tank is limited by
the already existing components. In the process of developing a fuel
tank the help of computer aided engineering (CAE) gets continuously
more attention. Although it will not fully replace real tests in the
near future, there are many possibilities where CAE can be of great
help. One option is the use of structural dynamics simulations where
(among other things) the behavior and durability of several parts
of the tank system under different conditions like changing pressure
or temperature can be examined. Another option is the usage of
computational fluid dynamics (CFD). An area where CFD can be of
help are sloshing simulations, which are needed for different reasons.
First there is the wish to get some insight into the behavior of the
liquid in a tank while the vehicle is driving. That means, given a
particular displacement or acceleration profile, analyzing the force
respectively pressure distribution and finding critical regions or the
built-in parts which are most likely the first to fail. Another point of
interest is the comparison of different shapes/designs with respect to
the fluid flow inside. A relatively new problem originates from the
concept of hybrid cars. In electrical mode they are very quiet and at
rapid velocity changes the sloshing noise from inside the tank can be
heard by the driver. By simulating the sloshing of fuel the reasons for
that sound emission can be identified and could be removed or at least
minimized with some design changes. There are many different things

© Springer Fachmedien Wiesbaden GmbH, part of Springer Nature 2019
M. Jäger, *Fuel Tank Sloshing Simulation Using the Finite Volume Method*,
BestMasters, https://doi.org/10.1007/978-3-658-25228-1_1

which have to be considered before doing an actual simulation. On one hand there is the necessary physical modeling for that kind of flow and on the other hand there is the numerical solution (approximation) method for that model. From the physical point of view the most important item is that more than one kind of fluid is involved. In most of the cases there is a liquid phase (e.g. fuel) and a gas phase (e.g. air). That means in addition to the flow in every phase there is an inner boundary between the phases which has to be modeled (the free surface). Moreover there will not always be only one connected liquid region, but several smaller parts. So the model should account for things like droplets breaking up or waves turning over. The two separated fluids are also one of the important things concerning the choice of numerical solution algorithm for the model, because this free surface and its evolution have to be computed. The approximation method should therefore be able to do this accurately enough. CFD could for example be done with mesh based methods like the finite volume method (FVM, see [16] or [34]), the finite element method (FEM, see [51] for a basic description and [18] for details on the application on fluid dynamics) or the finite difference method (FDM, shortly described in [16]). Independently of the choice between those methods there are several possibilities to describe the location and evolution of the free surface, like the Level Set (LS, proposed in 1988 in [38]) or Volume of Fluid method (VOF, originally described in 1981 by [23]). However, the need for a computational mesh is common to all of this methods. On contrary there are meshless methods, like the particle based Smoothed Particle Hydrodynamics (SPH,see [17] and [33]). In the SPH method the flow is, roughly speaking, computed by modeling the interaction between representative fluid particles in a small area around every particle. By doing so the free surface is automatically present through the distribution of the fluid particles. Most SPH implementations do parallelize very good and are running fast when used on GPUs. They are also accurate in predicting the free surface evolution ([45]). However, based on personal experience, there are inaccuracies in the evaluation of the pressure distribution on walls. This makes it difficult to use the method in conjunction with a structural simulation based on this pressure values. In this work the simulations will be based on a combination of FVM with a VOF approach for the free surface. As simulation tool the open source library OpenFOAM® (see [37]) was chosen. It is a very flexible framework for the description and solution of general systems of partial differential equations mainly based on the FVM. OpenFOAM® has many different features for a broad field of possible applications by default and allows the user to freely modify every part of the solver. Furthermore it has already been used for similar problems, including sloshing (see for example [31] or [9]).

2 Modeling

The term sloshing means any motion of a liquid inside a partially filled container and therefore including a free surface. Usually the sloshing is induced by the motion of the container, which often gets again influenced by the sloshing itself (e.g. damping). Due to the different physics involved, it is an unsteady, highly non-linear phenomenon and difficult to describe in full detail. Because of that, several simplifications have been used, both in the physical and the numerical model. The main assumption is that the continuum hypothesis should hold true. That means we are only interested in those problems where all the materials involved can be described as a continuum. The second very important assumption is that only the fluid flow is modeled. The tank walls are defined as a rigid body and no fluid-structure interaction is considered at that stage. As long as the resulting forces are not too strong and will not result in significant deformations of the tank walls there should not be too much of an error in the behavior of the fluids by neglecting this deformation.

A third major assumption is that no foam is considered in the model. Especially in the context of automotive fuels there is often a huge amount of foam created on the free surface. Nevertheless the creation of foam is completely ignored in the methodology used here. The reason for that is that the inclusion of an additional foam phase would lead to a significant more complex physical model which is out of the scope for this work. Other simplifications or models are concerning the characterization of the fluids or the flow and are described in the following sections alongside the basic equations used for describing free surface flows.

© Springer Fachmedien Wiesbaden GmbH, part of Springer Nature 2019
M. Jäger, *Fuel Tank Sloshing Simulation Using the Finite Volume Method*,
BestMasters, https://doi.org/10.1007/978-3-658-25228-1_2

2.1 General Equations

This section describes the partial differential equations used in this work. The specific notation and the necessary operators are given in Appendix 5. In a first approach every fluid is viewed as a separate domain. Therefore in every phase the conservation principles for a general fluid are valid (see [55] or [50] for a more detailed derivation). Expressed as partial differential equations the first principle is that of mass conservation

$$\frac{\partial \rho}{\partial t} + \nabla \cdot (\rho \boldsymbol{u}) = 0, \tag{2.1}$$

with the velocity $\boldsymbol{u}$ and the density ρ.
The second one is the linear momentum conservation

$$\frac{\partial \rho \boldsymbol{u}}{\partial t} + \nabla \cdot (\rho \boldsymbol{u} \boldsymbol{u}) - \nabla \cdot \boldsymbol{T} = \boldsymbol{f}, \tag{2.2}$$

where $\boldsymbol{T}$ is the stress tensor and $\boldsymbol{f}$ is the total body force (in most cases this is the gravitational force with $\boldsymbol{f} = \rho \boldsymbol{g}$). The stress tensor is symmetric due to the angular momentum conservation (see [50, §2.3]). The fluids considered here are all Newtonian fluids for which the stress tensor can be expressed as

$$\boldsymbol{T} = (-p + \lambda \nabla \cdot \boldsymbol{u})\boldsymbol{I} + 2\mu \boldsymbol{S}, \tag{2.3}$$

with the pressure p, the dynamic viscosity μ, the second coefficient of viscosity λ and the deformation tensor $\boldsymbol{S}$. Putting that into Equation (2.2) results in the momentum equation for Newtonian fluids

$$\frac{\partial \rho \boldsymbol{u}}{\partial t} + \nabla \cdot (\rho \boldsymbol{u} \boldsymbol{u}) + \nabla p - \nabla(\lambda \nabla \cdot \boldsymbol{u}) - \nabla \cdot (2\mu \boldsymbol{S}) = \boldsymbol{f}, \tag{2.4}$$

From Stokes' hypothesis (see [50, §3]) follows $\lambda = -\frac{2}{3}\mu$ and with the assumption of small deformations the deformation tensor can be written as

$$\boldsymbol{S} = \frac{1}{2}(\nabla\boldsymbol{u} + \nabla\boldsymbol{u}^T). \tag{2.5}$$

Combining all this, the momentum equation takes the form

$$\frac{\partial \rho\boldsymbol{u}}{\partial t} + \nabla \cdot (\rho\boldsymbol{u}\boldsymbol{u}) + \nabla p - \nabla \cdot \left[\mu \left((\nabla\boldsymbol{u} + \nabla\boldsymbol{u}^T) - \frac{2}{3}(\nabla \cdot \boldsymbol{u})\boldsymbol{I} \right) \right] = \boldsymbol{f}. \tag{2.6}$$

In this equation only the velocity, pressure and density fields are left as variables.

The third principle is that of energy conservation, written as

$$\frac{\partial}{\partial t}\rho\left(e + \frac{\boldsymbol{u}^2}{2}\right) + \nabla \cdot \left(\rho\left(e + \frac{\boldsymbol{u}^2}{2}\right)\boldsymbol{u}\right) - \nabla \cdot (\boldsymbol{T} \cdot \boldsymbol{u}) + \nabla \cdot \boldsymbol{q} = \boldsymbol{f} \cdot \boldsymbol{u}. \tag{2.7}$$

Here e means the internal energy and $\boldsymbol{q}$ the heat flux vector. By using the continuity and momentum equations this can be rewritten as

$$\rho(\frac{\partial e}{\partial t} + \boldsymbol{u} \cdot \nabla e) - \boldsymbol{T} : \nabla\boldsymbol{u} + \nabla \cdot \boldsymbol{q} = 0. \tag{2.8}$$

If we deal with Newtonian fluids and use Fourier's law

$$\boldsymbol{q} = -k\nabla T, \tag{2.9}$$

the energy equation takes the form

$$\rho(\frac{\partial e}{\partial t} + \boldsymbol{u} \cdot \nabla e) + p(\nabla \cdot \boldsymbol{u}) = \Phi + \nabla \cdot k\nabla T. \tag{2.10}$$

Here, T is the temperature, k the thermal conductivity and Φ the dissipation function, given by

$$\Phi = \lambda(\nabla \cdot \boldsymbol{u})^2 + 2\mu\boldsymbol{S} : \boldsymbol{S}. \tag{2.11}$$

If a fluid is assumed to be incompressible these equations can be simplified further. Note that incompressibility does not necessarily mean a constant density field, but that the density of a specific fluid particle does not change with time ([50]). In particular this can be described as a vanishing substantial derivative which is defined as

$$\frac{D}{Dt}(\cdot) := \frac{\partial(\cdot)}{\partial t} + \boldsymbol{u} \cdot \nabla(\cdot).$$

Therefore

$$\frac{D\rho}{Dt} = \frac{\partial\rho}{\partial t} + \boldsymbol{u} \cdot \nabla\rho = 0, \tag{2.12}$$

which reduces the continuity Equation (2.1) to the volume conservation

$$\nabla \cdot \boldsymbol{u} = 0. \tag{2.13}$$

That means for every particle the density is prescribed for all times through the initial conditions. Therefore Equation (2.4) can be written for incompressible fluids as

$$\frac{\partial\rho\boldsymbol{u}}{\partial t} + \nabla \cdot (\rho\boldsymbol{u}\boldsymbol{u}) = -\nabla p + \nabla \cdot \boldsymbol{\tau} + \boldsymbol{f}, \tag{2.14}$$

with the incompressible stress tensor $\boldsymbol{\tau} = 2\mu\boldsymbol{S}$. Using the additional assumption of a constant viscosity, the deformation tensor from Equation (2.5) and the identity $\nabla \cdot (\nabla\boldsymbol{u} + \nabla\boldsymbol{u}^T) = \nabla^2\boldsymbol{u}$, the momentum Equation (2.14) becomes

$$\rho\frac{\partial\boldsymbol{u}}{\partial t} + \rho\boldsymbol{u}\nabla\boldsymbol{u} + \nabla p - \mu\nabla^2\boldsymbol{u} = \boldsymbol{f}. \tag{2.15}$$

Sometimes this equation is divided by the density and the kinematic viscosity $\nu = \frac{\mu}{\rho}$ is introduced. That leads to

$$\frac{\partial \boldsymbol{u}}{\partial t} + \boldsymbol{u}\nabla\boldsymbol{u} + \frac{\nabla p}{\rho} - \nu\nabla^2\boldsymbol{u} = \frac{\boldsymbol{f}}{\rho}. \tag{2.16}$$

From equations (2.13) and (2.16) one can see that there are only $\boldsymbol{u}$ and p left as independent variables. Therefore, as long as no change in temperature occurs, the energy equation is not necessary for the solution of the flow field. In fluid dynamics this whole bunch of equations, namely the mass, momentum and energy equation, is often referred to as the Navier-Stokes equations.

2.2 Boundary and Interface Conditions

In all of the examined sloshing problems the whole domain is assumed to be closed without any inlet or outlet. So besides the interface only boundary conditions at the rigid tank walls are needed. The description of the necessary conditions follows in most parts the derivations in [55]. For the walls, there are generally two possible approaches: slip and no-slip conditions. No-slip means that the fluid particles stick to the wall and therefore they are having the same velocity

$$\boldsymbol{u} = \boldsymbol{u}_{wall}. \tag{2.17}$$

This condition resembles the physical situation better, whereas the slip condition is an approximation of this condition for small viscous stresses. A slip condition means that the fluid particles can freely slip along the wall and only their normal velocity is fixed. Therefore

$$\boldsymbol{u}_n = \boldsymbol{u}_{wall,n}, \tag{2.18}$$

$$\boldsymbol{u}_t - \boldsymbol{u}_{wall} = \beta\frac{\partial \boldsymbol{u}_t}{\partial \boldsymbol{n}}, \tag{2.19}$$

where β is a slip coefficient and the subscripts n and t are related to the normal and tangential directions. These two conditions are normally used dependent on the concrete problem. The no-slip boundary condition is for example suitable for viscous, incompressible fluids, whereas the slip condition can be used in the absence of viscous stresses. In this work we assume that the no-slip condition holds.

For the pressure in incompressible flows at a static wall usually the zero gradient boundary condition is applied. That means that the gradient perpendicular to the wall should vanish

$$\frac{\partial p}{\partial \boldsymbol{n}} = 0. \tag{2.20}$$

This condition actually is not really a physical boundary condition, because the incompressible flow model does not impose any condition on the pressure. In a numerical method it can be interpreted in a simple manner as a low order extrapolation, which will be shown in section 3.3.4. In a real-world application that condition has to be slightly modified so that the gradient is compatible with the tangential velocity and the velocity gradient at the wall [34, §15.6]. Especially for moving walls it has to be adjusted to accomplish for the additional wall velocity. Sloshing is mostly induced by the displacement of the tank under consideration. This excitation can be applied to the model in two different ways. One is the application of a rigid body motion to the tank geometry. This results in a non-zero velocity at the walls and is therefore included in the boundary conditions. The other option is to do the simulation using a non-inertial frame of reference ([50]). This means that the equations are formulated for an observer which is moving with the tank. The tank displacement is then applied through additional source terms in the momentum equation ([44]).

In the two fluid approach, where every phase is treated as a separate domain, the interface forms an additional boundary Γ. Therefore additional boundary (or interface) conditions are needed. In the following derivations it is assumed that the fluids are Newtonian, incompressible and that no phase change respectively mass flow between the phases occurs. Real fluids will not fully satisfy this simplification. However, under several conditions they can be treated as if they were like that (see §2.4). The different fluids are generally named as fluid 1 (the reference fluid, in most cases the liquid) and fluid 2. For any quantity ϕ the subscript ϕ_k means here the respective quantity of the k-th phase (e.g. the viscosities μ_1 and μ_2). The interface conditions are then formulated as

$$\boldsymbol{u}_1 = \boldsymbol{u}_2, \qquad (2.21)$$

$$-\left(-p + 2\mu\boldsymbol{n} \cdot \boldsymbol{S} \cdot \boldsymbol{n}\right)_2 + \left(-p + 2\mu\boldsymbol{n} \cdot \boldsymbol{S} \cdot \boldsymbol{n}\right)_1 = \sigma\kappa, \qquad (2.22)$$

$$-\left(2\mu\boldsymbol{t}^{(l)} \cdot \boldsymbol{S} \cdot \boldsymbol{n}\right)_2 + \left(2\mu\boldsymbol{t}^{(l)} \cdot \boldsymbol{S} \cdot \boldsymbol{n}\right)_1 = \boldsymbol{t}^{(l)} \cdot \nabla_\Gamma\sigma. \qquad (2.23)$$

In these equations $\boldsymbol{n}$ is the unit normal vector directed from fluid 1 to fluid 2 and $\boldsymbol{t}^{(l)}$ ($l = 1, 2$) are unit tangent vectors at the interface. Furthermore σ means the surface tension coefficient and κ the interface curvature. The ∇_Γ symbol is the surface gradient (see [55, §2.4.2 and Appendix A]). The first equation assures continuity of the velocity across the interface. Additionally the second and third equations are the jump conditions for stresses in normal and tangential directions (jump condition because of the discontinuity due to surface tension). The evaluation of those relations, especially the surface curvature, requires an accurate description of the position of the interface. This can be done in several ways. One is the parametrized description as a surface (or separate surfaces for every region) in the three dimensional space. This is represented by a mapping

$$\boldsymbol{X}(u, v) = (x(u, v), y(u, v), z(u, v)), \qquad (2.24)$$

with two independent parameters u and v (see also [55, §2.3]). Another possibility is the description through a phase function. Here two different methods to define this phase function are given. The first is the definition of the interface as an iso-surface through a specific value of a smooth function F. By using the 0 contour for the interface the different fluid regions are then defined by the values $F > 0$ or $F < 0$. With $F > 0$ as the reference phase the outwards pointing surface normal is given by

$$n = -\frac{\nabla F}{|\nabla F|} \tag{2.25}$$

and the surface curvature by

$$\kappa = -\nabla \cdot n = \nabla \cdot \left(\frac{\nabla F}{|\nabla F|}\right). \tag{2.26}$$

This approach is followed by the level-set methods ([55, §4.5]). The other approach is by using the characteristic function $H(x)$ which takes a value of 1 in the area of fluid 1 and 0 in the other regions. In this case the fluid 1 was taken as the reference phase, but this can also be done the other way around or with multiple different fluids where each of them has its own phase function. The interface is then described by this discontinuity. In that approach one has to pay attention to the fact that this function is not differentiable in the conventional sense, although the gradient is needed for the surface curvature. Therefore the concept of distributions is used, where the derivative of a characteristic function can be described with a δ-distribution (for details see [55, §2.3 or Appendix A]). The position of the interface changes with time. Nevertheless, every single fluid particle has its own value for H which does not change with time, defined through the initial or boundary conditions. Therefore the material derivative of H is zero leading to the equation

$$\frac{\partial H}{\partial t} + u \cdot \nabla H = 0 \tag{2.27}$$

for the evolution of the phase function ([55, §4.1]).

2.3 One Fluid Formulation

The whole free surface model can also be viewed in a so called one fluid formulation, where the different phases are treated as one fluid with varying properties, depending on the respective phase. This approach is especially interesting with respect to the numerical procedure described in Section 3.4. In the one fluid approach the properties of both fluids are combined to a mixture by using the phase function H from Section 2.2 to identify the different regions. Taking the densities ρ_1 and ρ_2 as an example the mixture density is then defined as

$$\rho = H\rho_1 + (1 - H)\rho_2. \tag{2.28}$$

This mixture is then also using only one pressure field and one mixture velocity.

The formulation of the incompressible continuity equation (Eq. (2.13)) is not affected by that mixture model, because the separate velocities are just replaced by the mixture velocity. The same goes for the momentum equation, but it needs to be further modified to account for the additional interface physics like the surface tension. In our case that is added to the equation as an additional surface body force $\boldsymbol{f}_\sigma \delta_\Gamma$ following the Continuum Surface Force (**CSF**) model ([7]). Here $\delta_\Gamma = \delta(\boldsymbol{x} - \boldsymbol{x}_\Gamma)$ is the Dirac-delta distribution with respect to the interface coordinates $\boldsymbol{x}_\Gamma$ ([55, Appendix A]). Therefore (2.16) changes to

$$\rho\frac{\partial \boldsymbol{u}}{\partial t} + \rho\boldsymbol{u}\nabla\boldsymbol{u} + \nabla p - \mu\nabla^2\boldsymbol{u} = \boldsymbol{f} + \boldsymbol{f}_\sigma\delta_\Gamma. \tag{2.29}$$

In the case of constant surface tension $\boldsymbol{f}_\sigma$ can be rewritten as

$$\boldsymbol{f}_\sigma\delta_\Gamma = \sigma\kappa\boldsymbol{n}\delta_\Gamma$$

with σ and κ according to Equation (2.22).

So far only the situation of an existing interface moving with time got addressed, but nothing has been said about possible topological changes. Such changes like the break up or coalescence of droplets are very common in free surface flows. According to [55] this situations can generally be reduced to two situations. Those are the rupture of thin films and the snapping of thin threads in one of the fluid areas. In these two situations molecular effects are always present in reality, but they can not be described in the framework of the continuum hypotheses. Details on slightly bigger scale, on the other hand, can be treated with continuum mechanics. The snapping threads can be described accurately enough by the standard Navier-Stokes equations as their diameter becoming zero in finite time should be accounted for without modifying the equations ([55, 2.7]). The problem of rupturing films on the other hand needs some special care. The intermolecular forces acting on a thin film with a thickness less then a few hundred nanometers can be modeled as a singular surface force ([55, §2.7.1]). This force per unit surface area, whose direction is away from the opposite interfaces, can be taken as

$$f_I = -Ah^{-3},\tag{2.30}$$

where h means the distance between the two interfaces (see image 2.1) and A is the Hamaker constant.

The surface force in (2.29) is then given by

$$f_\sigma \delta_\Gamma = -Ah^{-3}n\delta_\Gamma + \sigma\kappa n\delta_\Gamma + \nabla_\Gamma\sigma\delta_\Gamma.\tag{2.31}$$

In this equations the value of A is fluid dependent, but generally a value of $A > 0$ results in interfaces which attract each other, as it is the case for pure water, while $A < 0$ means the opposite (see [55, §2.7.1]). Although the surface force from Equation (2.31) can be used in solving the macroscopic Navier-Stokes equations, it is in many cases not really efficient to do so. The needed scales in time and space are so small that the computational afford to resolve them in

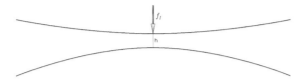

Figure 2.1: Surface force at thin film

the numerical solution procedure would be too high. As the influence of this additional force is usually only important at those very small scales, it is mostly sufficient to just use the Navier-Stokes equations in a formulation where this force f_I is simply neglected.

2.4 Incompressibility

Under the assumption of immiscible, incompressible and isothermal fluids the physics of a general sloshing model can be described by the equations which are presented so far. That leads to the question whether this constraints are valid and under which conditions. Generally a fuel tank is exposed to major changes in temperature. However, most of the sloshing takes place at small time scales and nearly constant environmental conditions, which in this case means also nearly constant temperature. Therefore an isothermal model should be a good approximation of reality if only the effects of sloshing are considered. Whether the fluids are miscible or not depends on their properties, but combinations like water-air or diesel-air at normal room temperature, which are mostly used in this work will fulfill this condition. The remaining condition of incompressibility is a bit more difficult to justify. That is because this depends not only on the material and thermophysical properties, but is related to the characteristics of the flow. In reality incompressible fluids do not exist. However, for applications of sloshing in fuel tanks and under the assumptions made so far, it is sufficient to assume incompressibility for fluids in liquid form. Although gases, especially air, are truly compressible there are conditions where they can be modeled as incompressible fluids. According to [50], several characteristic numbers are suitable for confirming that. For that we define U as a characteristic velocity, L as a typical length, a as the speed of sound and f as a typical frequency (which also means $\tau := f^{-1}$ as a characteristic time interval). The conditions are then written as

$$\frac{gL}{a^2} \ll 1, \tag{2.32}$$

$$\frac{U^2}{a^2} = M \ll 1, \tag{2.33}$$

$$\frac{L^2 f^2}{a^2} \ll 1 \text{ or } \frac{L}{a} \ll \tau. \tag{2.34}$$

The condition on the Mach number M in (2.33) is often taken to be the most important number. Many take $M < 0.3$ or 0.2 as a sufficient condition for incompressibility (e.g. [16], [34]). So in order to use the incompressible model as described before, those conditions have to be satisfied, especially in the gaseous area. Therefore this justification has to be done separately for any specific problem. Only some general facts can be given here. The first observation is that the speed of sound in air at $25°C$ is about $346\frac{m}{s}$ ([1, §4.4.3]). Because sound is traveling the slowest in gaseous media this can be taken as some kind of lower bound for the real value. Therefore the term $\frac{a^2}{g}$ will be at least of order 10^5. That means that condition (2.32) should always be true in those dimensions where the sloshing takes place. The same goes for condition (2.34) as long as the main sloshing frequencies are not too high. That leaves the Mach number condition (2.33), depending on a characteristic velocity, as the most important point to verify.

2.5 Turbulence

Generally the Navier-Stokes equations completely describe the whole range of fluid flows from laminar to fully turbulent, as long as the continuum hypothesis holds true. However, in numerical solution procedures it is in most cases not even possible to fully resolve every detail of turbulence. So one has to distinguish between different levels of turbulence modeling (see [16]). The most detailed version is the Direct Numerical Simulation (DNS) where all of the motions down to the Kolmogorov scales (see [59]) are resolved, which makes this approach computationally very expensive. In the Large Eddy Simulation (LES) only the large scales (those which are resolvable by the used grid) are computed exactly whereas the smaller scales are modeled by a so called subgrid-scale (SGS) model. Although it is computationally cheaper than a DNS, in most cases this approach is also too expensive to be of practical use. As in many fluid problems one is not interested in all of the small and random turbulent fluctuations, but in general (or average) properties of the flow, the use of Reynolds-Averaged Navier-Stokes (RANS) equations is often sufficient. If the inclusion of turbulence is necessary this approach is also followed in this work and therefore will be shortly described here. In RANS simulations all the computed quantities are statistically averaged (see [16, §9] or [34, §17]). In this model any quantity ϕ is split into an average value and its fluctuation around that value

$$\phi(\boldsymbol{x}, t) = \overline{\phi}(\boldsymbol{x}, t) + \phi'(\boldsymbol{x}, t).$$

The definition of the average for a given quantity ϕ depends on the problem. For statistically steady flows the average can be computed as the time average

$$\overline{\phi}(\boldsymbol{x}) = \lim_{T \to \infty} \frac{1}{T} \int_0^T \phi(\boldsymbol{x}, t) \, \mathrm{d}t.$$

In unsteady flows this average is replaced by the ensemble average, where an ensemble is a big enough number of realizations of the same flow situation. Then the average is defined as

$$\overline{\phi}(\boldsymbol{x}, t) = \lim_{N \to \infty} \frac{1}{N} \sum_{n=1}^{N} \phi^{(n)}(\boldsymbol{x}, t),$$

with N the number of realizations in the ensemble. Independent of the choice of averaging process the term "Reynolds averaging" refers to any of them. To derive the RANS equations all of the averaging strategies have to fulfill the following rules (ϕ and ψ are general variables):

$$\overline{\phi'} = 0,$$
$$\overline{\overline{\phi}} = \overline{\phi},$$
$$\overline{\nabla\phi} = \nabla\overline{\phi},$$
$$\overline{\phi + \psi} = \overline{\phi} + \overline{\psi},$$
$$\overline{\overline{\phi}\psi} = \overline{\phi}\overline{\psi},$$
$$\overline{\overline{\phi}\psi'} = 0,$$
$$\overline{\phi\psi} = \overline{\phi}\overline{\psi} + \overline{\phi'\psi'}.$$

With that in mind the Navier-Stokes equations are rewritten with every quantity split into an average part and a fluctuating part. Then every equation gets "Reynolds averaged". For the case of incompressible Newtonian fluids, governed by the equations

$$\nabla \cdot (\rho \boldsymbol{u}) = 0,$$
$$\frac{\partial \rho \boldsymbol{u}}{\partial t} + \nabla \cdot (\rho \boldsymbol{u} \boldsymbol{u}) = -\nabla p + \nabla \cdot \boldsymbol{\tau} + \boldsymbol{f},$$

and using a general stress tensor $\boldsymbol{\tau}$ and body force $\boldsymbol{f}$, this results in the equations

$$\nabla \cdot (\rho \bar{\boldsymbol{u}}) = 0, \tag{2.35}$$

$$\frac{\partial \rho \bar{\boldsymbol{u}}}{\partial t} + \nabla \cdot (\rho \overline{\boldsymbol{u}\boldsymbol{u}}) = -\nabla \bar{p} + \nabla \cdot (\bar{\boldsymbol{\tau}} - \rho \overline{\boldsymbol{u}'\boldsymbol{u}'}) + \bar{\boldsymbol{f}}. \tag{2.36}$$

These are very similar to the original equations except for the additional tensor $\boldsymbol{\tau}^R := -\rho \overline{\boldsymbol{u}'\boldsymbol{u}'}$, which is known as the Reynolds stress tensor and introduces 6 additional unknowns. That leaves less equations than unknowns and makes this set of equations non closed. A turbulence model is an approach to close the set by expressing the Reynolds stresses in terms of the mean values. That is often done analogous to the stress tensor in Newtonian flows by expressing the Reynolds stresses as a linear function of the mean velocity gradients. That means

$$\boldsymbol{\tau}^R = -\rho \overline{\boldsymbol{u}'\boldsymbol{u}'} := \mu_t \left[\nabla \bar{\boldsymbol{u}} + (\nabla \bar{\boldsymbol{u}})^T \right] - \frac{2}{3} \left[\rho k + \mu_t (\nabla \cdot \bar{\boldsymbol{u}}) \right] \boldsymbol{I} \tag{2.37}$$

with the turbulent kinetic energy

$$k = \frac{1}{2} \overline{\boldsymbol{u}' \cdot \boldsymbol{u}'} \tag{2.38}$$

and the turbulent eddy viscosity μ_t. This assumption is known as the Boussinesq Hypothesis ([5],[34]). In the models based on this approximation the calculation of the Reynolds stresses is done through the calculation of turbulent kinetic energy and turbulent viscosity. Those models are generally grouped into categories corresponding to the numbers of additional transport equations which have to be solved [16]. Here the $k - \varepsilon$ and the $k - \omega$ model are presented (see [59]), which are both so called two-equation models. The $k - \varepsilon$ model ([26]) is based on the equation

$$\mu_t = \rho C_\mu \frac{k^2}{\varepsilon} \tag{2.39}$$

for the turbulent viscosity, where ε is the rate of dissipation of turbulence kinetic energy per unit mass due to viscous stresses defined as

$$\varepsilon = \frac{1}{2}\frac{\mu}{\rho}\overline{\{\nabla u' + (\nabla u')^T\} : \{\nabla u' + (\nabla u')^T\}}. \tag{2.40}$$

The necessary transport equations for k and ε are

$$\frac{\partial(\rho k)}{\partial t} + \nabla \cdot (\rho u k) = \nabla \cdot (\mu_{eff,k}\nabla k) + P_k - \rho\varepsilon \tag{2.41}$$

and

$$\frac{\partial(\rho\varepsilon)}{\partial t} + \nabla \cdot (\rho u\varepsilon) = \nabla \cdot (\mu_{eff,\varepsilon}\nabla\varepsilon) + C_{\varepsilon 1}\frac{\varepsilon}{k}P_k - C_{\varepsilon 2}\rho\frac{\varepsilon^2}{k}, \tag{2.42}$$

where P_k is the production of turbulent energy defined as

$$P_k = \tau^R : \nabla u. \tag{2.43}$$

In these equations the viscosities are defined as

$$\mu_{eff,k} = \mu + \frac{\mu_t}{\sigma_k} \text{ and } \mu_{eff,\varepsilon} = \mu + \frac{\mu_t}{\sigma_\varepsilon}. \tag{2.44}$$

The constants in these equations are generally flow dependent, but in most cases they are defined as $C_{\varepsilon 1} = 1.44$, $C_{\varepsilon 2} = 0.09$, $\sigma_k = 1.0$ and $\sigma_\varepsilon = 1.3$. The $k - \omega$-model ([58],[59]), also based on the Boussinesq approximation, introduces another variable instead of ε. It is based on the specific turbulence dissipation (=rate at which turbulence kinetic energy is converted into internal thermal energy per unit volume and time)

$$\omega = \frac{\varepsilon}{C_\mu k} \tag{2.45}$$

and the turbulent viscosity

$$\mu_t = \rho \frac{k}{\omega}. \tag{2.46}$$

The respective transport equations are given by

$$\frac{\partial(\rho k)}{\partial t} + \nabla \cdot (\rho \boldsymbol{u} k) = \nabla \cdot (\mu_{eff,k} \nabla k) + P_k - \beta^* \rho k \omega \tag{2.47}$$

and

$$\frac{\partial(\rho \omega)}{\partial t} + \nabla \cdot (\rho \boldsymbol{u} \omega) = \nabla \cdot (\mu_{eff,\omega} \nabla \omega) + C_{\alpha 1} \frac{\omega}{k} P_k - C_{\beta 1} \rho \omega^2. \tag{2.48}$$

In these equations the viscosities are

$$\mu_{eff,k} = \mu + \frac{\mu_t}{\sigma_{k1}} \text{ and } \mu_{eff,\omega} = \mu + \frac{\mu_t}{\sigma_{\omega 1}}, \tag{2.49}$$

with the constants often defined as $C_{\alpha 1} = \frac{5}{9}$, $C_{\beta 1} = 0.075$, $\beta^* = 0.09$, $\sigma_{k1} = 2$ and $\sigma_{\omega 1} = 2$.
While the $k - \varepsilon$ model is performing good for free-shear flows it is having problems in predicting the flow in regions with adverse pressure gradient. The $k - \omega$ model on the other hand is more robust, directly applicable through the sub-layer and better suitable for adverse pressure gradients. However, it is very critical to define the free-stream conditions in the right way (see [34, §1.7.2]). Therefore another model that is regularly used has been developed. This is the $k - \omega - SST$ model (see [32]), which is based both on the $k - \varepsilon$ and the $k - \omega$ model with the aim of combining the advantages of both models. For

simplicity reasons it will not be described here.

Each of these models has its one advantages and drawbacks and by using them, one has to always keep in mind that they all introduce additional modeling compared to the original Navier-Stokes equations. Additionally, when using such a turbulence model, it is always necessary to pay attention to the flow at near wall regions by using the right boundary conditions and wall functions in combination with a suitable mesh. Some details on the models, wall functions and their implementation in finite volume methods can be found in [34, §17].

2.6 Summary of Equations

The complete two-phase model used in this work is now described by the equations for incompressible immiscible isothermal Newtonian fluids with a free surface as explained in the previous sections. This means primarily the incompressible Navier-Stokes Equations given by the mass and linear momentum conservation equations

$$\nabla \cdot \boldsymbol{u} = 0,$$

$$\rho \frac{\partial \boldsymbol{u}}{\partial t} + \rho \boldsymbol{u} \nabla \boldsymbol{u} = -\nabla p + \mu \nabla^2 \boldsymbol{u} + \boldsymbol{f}. \qquad \text{(N.-S.)}$$

These are combined with the phase fraction equation

$$\frac{\partial H}{\partial t} + \boldsymbol{u} \cdot \nabla H = 0 \qquad (2.50)$$

and the mixture properties

$$\rho = H\rho_1 + (1 - H)\rho_2, \qquad (2.51)$$

$$\mu = H\mu_1 + (1 - H)\mu_2. \qquad (2.52)$$

Dependent on the used models for turbulence and surface tension the momentum equations have to be slightly modified (mainly the effective viscosity and the source terms will be changed) and eventually some additional turbulence equations are to be solved.

3 Discretization

In this section the solution procedure for the system of equations derived in the last section will be addressed. It starts with a short comment on the existence and uniqueness of a solution for the incompressible Navier-Stokes equations given by the equations (N.-S.). The Finite Volume Method will be presented in several steps. First the discretization of a general transport equation with the FVM will be described. Based on these techniques the methodology for the numerical solution of the Navier-Stokes equations is explained. After that the representation and advection of the free surface on a discrete mesh with the VOF method will be discussed concluding with the integration of the VOF method in the solution algorithm. At the end of this section a short summary of the algorithm used will be given.

3.1 Existence and Uniqueness

Before applying any solution method on any system of equations, one has to consider whether a unique solution of this system does even exist. In the case of the Navier-Stokes equations there are some problems in this context ([54], [6]). The solvability of these equations is often examined using a general weak formulation and therefore searching for a unique weak solution. A weak formulation of the Navier-Stokes equations can be found in [54, Ch. III, Problem 3.2]. For dimension $n \leq 4$ there exists a weak solution to the weak problem in a specific function space depending on the regularity of the boundary conditions and source terms (see. [54, Ch. III, Theorem 3.1]). In the case of a 2-dimensional space also the uniqueness of this solution in the same function space can be established (see. [54, Ch. III, Theorem 3.2]). However for the 3-dimensional space this

© Springer Fachmedien Wiesbaden GmbH, part of Springer Nature 2019
M. Jäger, *Fuel Tank Sloshing Simulation Using the Finite Volume Method*,
BestMasters, https://doi.org/10.1007/978-3-658-25228-1_3

can not be shown in the same way. There are results which give a
unique solution, but this solution is not lying in the same function
space as used in the existence theorems. In fact it can be shown that
there exists at most one solution, but in a space which is smaller
(see. [54, Ch. III, Theorem 3.4]). This problem for the 3D general
Navier-Stokes equations is still unsolved at the moment. However,
there are some results showing both existence and uniqueness under
specific a priori estimates respectively conditions (see [6, Ch. 5]).
Nevertheless, for the derivation of the FVM method, it is necessary
to assume that there is a smooth enough solution so that for example
the Taylor Series Expansion Theorem is applicable. That has to be
kept in mind when thinking about accuracy or convergence.

3.2 The Finite Volume Method

The finite volume method is explained according to [25] and [34].
Since the basic equations from section 2 are all some kind of transport
equation it suffices to consider only the general transport equation
for any scalar property $\phi(\boldsymbol{x}, t)$,

$$\underbrace{\frac{\partial \rho \phi}{\partial t}}_{\text{I}} + \underbrace{\nabla \cdot (\rho \boldsymbol{u} \phi)}_{\text{II}} - \underbrace{\nabla \cdot (\rho \Gamma_\phi \nabla \phi)}_{\text{III}} = \underbrace{S_\phi(\phi)}_{\text{IV}}, \qquad (3.1)$$

depending on a given density ρ and velocity field $\boldsymbol{u}$. The first term
(I) is the temporal derivative which expresses the change of ϕ with
time at any point in space. The second term (II) is the convection
term which generally describes the change of ϕ by material transport
through the velocity field. The third term (III) is the diffusion term
with the diffusivity Γ_ϕ. It means the change of ϕ due to its own
gradient (for example change in concentration of some transported
quantity). The $S_\phi(\phi)$ term (IV) is the source term, which includes
every other effect on ϕ. That could for example be some external
forces or some effects which create sources or sinks. Note that in
general the source term can also depend on ϕ itself. At some points

in the following derivations a modified one-dimensional formulation of this equation will be used to give better insight in particular aspects of the discretization. This equation is given by

$$\frac{\partial \rho \phi}{\partial t} + \frac{\partial \rho u \phi}{\partial x} - \frac{\partial}{\partial x}(\rho \Gamma_\phi \frac{\partial}{\partial x}\phi) = S_\phi. \tag{3.2}$$

Any method for the discretization of such an equation can be evaluated with respect to a few basic criteria (see [34, §5.6]). Some of the most important criteria are

- **Conservation**
  That means that any transported property should be preserved throughout the computational domain. It basically says that the flux leaving one cell through a face has to be equal to the flux entering the neighboring cell through the same face. This property is strongly dependent on the formulation of the conservation principles (conservative vs non-conservative formulation, see [22]).

- **Accuracy**
  This property describes the error between the approximated and the exact solution. As the exact solution is in most cases not known, an alternative is to look at the truncation error of a suitable taylor expansion series. Although this value can not really give informations about the absolute error it can give some insight into the rate at which the error will decrease with further grid refinement.

- **Convergence**
  This term can be used in several ways. In general, a method has converged if a solution is obtained. However, it can also mean that a solution does not change with further grid refinement or by marching forward in time. In iterative methods it says that the change between consecutive iterations or the residual is smaller than a predefined value.

- **Consistency**
  The method is said to be consistent if approximate solutions are

approaching the exact solution at any time or point in space if the grid size and time step go to zero. Particularly, this means that the discretization error has to depend on positive powers of both Δt and $\Delta \boldsymbol{x}$.

- **Stability**
  For transient problems a method is said to be stable if errors are kept bounded when marching forward in time. Additionally, it can also describe how an iterative solution method for the discretized equations reacts to errors or how variations in the initial and boundary conditions affect the solution.

- **Transportiveness**
  This means how the discretization method depicts the transport characteristics of the original system of equations. In particular we refer to the ratio of convective to diffusive transport which is measured by the Péclet number.

$$Pe = \frac{Convection}{Diffusion} = \frac{\rho u}{\Gamma_\phi / \Delta x} \tag{3.3}$$

where $Pe = 0$ indicates pure diffusion. Here the discrete equations should exhibit the same behavior as the continuous equations

- **Boundedness**
  By boundedness one refers to the ability of the method to prevent unphysical over- or undershoots. That means for example that when no sources or sinks exist the value of a conserved quantity at any point in the domain should be bounded by the initial, boundary or neighboring values of that point.

A Finite Volume Method can be constructed in different ways where the approximate solution is then described by its values at the defined computational nodes. It could be introduced using the vertices of a mesh as computational nodes which results in a so-called vertex-based FVM or it can be constructed based on the cells of a mesh. The method described in this work is cell-based and applicable to any unstructured mesh. That means the computational domain can be split into cells/control volumes of any polyhedral shape. The only restriction on the mesh is that all the cells have to be convex and bounded by flat faces. This limitation has theoretical reasons and assures the validity of the needed mathematical theorems (see Appendix 5). However, modern FVM codes are often able to give acceptable results even on meshes with non-flat faces as mentioned in [44]. The computational nodes are the midpoints of the cells or control volumes. For a cell V_P the midpoint $\boldsymbol{x}_P$ is defined by the equation

$$\int_{V_P} (\boldsymbol{x} - \boldsymbol{x}_P) \, \mathrm{d}\boldsymbol{x} = 0. \tag{3.4}$$

Because of the assumed convexity of the cells this midpoint is lying inside the control volume. The bounding faces with area S are referred to as f and for each of them the face midpoint is given as $\boldsymbol{x}_f$, which is defined analogously to (3.4) as

$$\int_{S_f} (\boldsymbol{x} - \boldsymbol{x}_f) \, \mathrm{d}s_{\boldsymbol{x}} = 0. \tag{3.5}$$

The idea behind the finite volume method is to assure that Equation (3.1) should be valid in every control volume. That means given an arbitrary control volume V_P, a time instant t and time step Δt the equation has to be satisfied for all points in the control volume. Integrating this equation first in space over V_p and then in time from t to $t + \Delta t$ gives the integral conservation equation

$$\int_t^{t+\Delta t} \left(\int_{V_P} \frac{\partial}{\partial t} (\rho \phi) \, \mathrm{d}\boldsymbol{x} + \int_{V_P} \nabla \cdot (\rho \boldsymbol{u} \phi) \, \mathrm{d}\boldsymbol{x} \right) \mathrm{d}t$$
$$- \int_t^{t+\Delta t} \left(\int_{V_P} \nabla \cdot (\rho \Gamma_\phi \nabla \phi) \, \mathrm{d}\boldsymbol{x} \right) \mathrm{d}t = \int_t^{t+\Delta t} \left(\int_{V_P} S_\phi(\phi) \, \mathrm{d}\boldsymbol{x} \right) \mathrm{d}t. \tag{3.6}$$

Equation (3.6) is used as the base for the finite volume method, where every term of this equation needs to be discretized separately. The quality of the discretization depends on the assumed variation of ϕ over any control volume.

In this case the variation is restricted to be linear in both space and time in every cell. Therefore a Taylor series expansion of ϕ in space and time is used, where the terms with order higher than 2 are neglected (see Appendix 5, Theorem .1).

The control volumes are assumed to be compact subsets of $\mathbb{R}^d$ by construction. That means for every control volume V_P there exists an open set $\Omega_P \supset V_P$. For every $\boldsymbol{x} \in V_P \subset \Omega_P$ it holds true that also the path to $\boldsymbol{x}_P$ is in V_P. If additionally, $\phi(\boldsymbol{x}, t)$ is at least $C^2(\Omega_P)$ for any fixed time t, it can be written according to

$$\phi(\boldsymbol{x}, t) = \phi(\boldsymbol{x}_P, t) + (\boldsymbol{x} - \boldsymbol{x}_P) \cdot \nabla\phi(\boldsymbol{x}_P, t) + \mathcal{O}(|\boldsymbol{x} - \boldsymbol{x}_P|^2) \qquad (3.7)$$

In time it is necessary to use the one dimensional version of Theorem .1 (see for example [21, §VIII]) with the fact that for a finite Δt any interval $[t, t + \Delta t]$ is also compact. That means for every time interval there exists an open set $\Omega_t \supset [t, t + \Delta t]$. Let $\boldsymbol{x}$ now be arbitrary but fixed. Then, if $\phi(\boldsymbol{x}, t) \in C^2(\Omega_t)$, it can be written similar to (3.7) as

$$\phi(\boldsymbol{x}, t + \Delta t) = \phi(\boldsymbol{x}, t) + \Delta t \frac{\partial \phi(\boldsymbol{x}, t)}{\partial t} + \mathcal{O}((\Delta t)^2). \qquad (3.8)$$

For a second order accurate finite volume method those last terms are simply neglected. That means if Δt and $\Delta \boldsymbol{x}$ are small enough the variation of ϕ in a control volume (both in space and time) is then expressed as

$$\phi(\boldsymbol{x}, t) \approx \phi_P + (\boldsymbol{x} - \boldsymbol{x}_P) \cdot (\nabla\phi)_P \qquad (3.9)$$

$$\phi(\boldsymbol{x}, t + \Delta t) \approx \phi^t + \Delta t \left(\frac{\partial \phi}{\partial t}\right)^t. \qquad (3.10)$$

The sub- and superscripts are used for better readability and will be regularly used from here on. They mean the evaluation of ϕ at certain discrete points and are defined as

$$\phi_P = \phi(\boldsymbol{x}_P) \qquad\qquad \text{(cell center)}, \qquad (3.11)$$

$$\phi_f = \phi(\boldsymbol{x}_f) \qquad\qquad \text{(face center)}, \qquad (3.12)$$

$$\phi^n = \phi(t_n) \qquad \text{(discrete time instance } t_n\text{)}. \qquad (3.13)$$

Taking the integral of ϕ over V_P and using equations (3.9) and (3.4) leads to

$$\int_{V_P} \phi(\boldsymbol{x}, t) \, \mathrm{d}\boldsymbol{x} \stackrel{(3.9)}{\approx} \int_{V_P} [\phi_P(t) + (\boldsymbol{x} - \boldsymbol{x}_P) \cdot (\nabla\phi(t))_P] \, \mathrm{d}\boldsymbol{x}$$

$$= \phi_P(t) \int_{V_P} \mathrm{d}\boldsymbol{x} + \underbrace{\left[\int_{V_P} (\boldsymbol{x} - \boldsymbol{x}_P) \, \mathrm{d}\boldsymbol{x} \right]}_{=0\ (3.4)} \cdot (\nabla\phi(t))_P$$

$$= \phi_P(t) \, |V_P| . \qquad (3.14)$$

Looking at this the other way around it is in fact a mean value approximation where the mean value is defined as

$$\overline{\phi}_P(t) = \frac{\int_{V_P} \phi(\boldsymbol{x}, t) \, \mathrm{d}\boldsymbol{x}}{|V_P|}$$

$$\stackrel{(3.9)}{=} \frac{1}{|V_P|} \int_{V_P} [\phi_P(t) + (\boldsymbol{x} - \boldsymbol{x}_P) \cdot (\nabla\phi(t))_P + \mathcal{O}(|\boldsymbol{x} - \boldsymbol{x}_P|^2)] \, \mathrm{d}\boldsymbol{x}$$

$$= \phi_P(t) + \mathcal{O}(|\boldsymbol{x} - \boldsymbol{x}_P|^2). \qquad (3.15)$$

This is therefore an approximation of the value at the midpoint with second order accuracy.

In the following derivations the Gauss or Divergence Theorem will be needed (see Appendix 5, Theorem .2). Let $\boldsymbol{a}$ be an arbitrary vector field which is assumed to vary linearly over each control volume. Since any control volume V_P is compact and bounded by flat and therefore smooth faces, Gauss' divergence theorem can be used to derive the following identities

$$\int_{V_P} \nabla \cdot \boldsymbol{a} \, \mathrm{d}\boldsymbol{x} = \oint_{\partial V_P} \boldsymbol{a} \cdot \boldsymbol{n} \, \mathrm{d}s_{\boldsymbol{x}} \tag{3.16}$$

$$\int_{V_P} \nabla \phi \, \mathrm{d}\boldsymbol{x} = \oint_{\partial V_P} \phi \boldsymbol{n} \, \mathrm{d}s_{\boldsymbol{x}} \tag{3.17}$$

$$\int_{V_P} \nabla \boldsymbol{a} \, \mathrm{d}\boldsymbol{x} = \oint_{\partial V_P} \boldsymbol{a} \boldsymbol{n} \, \mathrm{d}s_{\boldsymbol{x}}. \tag{3.18}$$

With the first equation follows

$$\int_{V_P} \nabla \cdot \boldsymbol{a} \, \mathrm{d}\boldsymbol{x} = \oint_{\partial V_P} \boldsymbol{a} \cdot \boldsymbol{n} \, \mathrm{d}s_{\boldsymbol{x}} = \sum_f \left(\int_{S_f} \boldsymbol{a} \cdot \boldsymbol{n} \, \mathrm{d}s_{\boldsymbol{x}} \right).$$

Using the linear variation of $\boldsymbol{a}$ over a face f for every component a_i one gets

$$\int_{S_f} a_i n_i \, \mathrm{d}s_{\boldsymbol{x}} = \int_{S_f} [(a_i)_f + (\boldsymbol{x} - \boldsymbol{x}_f) \cdot (\nabla a_i)_f] \, n_i \, \mathrm{d}s_{\boldsymbol{x}}$$

$$\approx (a_i)_f n_i \underbrace{\int_{S_f} \mathrm{d}s_{\boldsymbol{x}}}_{=|S_f|} + \underbrace{\left[\int_{S_f} (\boldsymbol{x} - \boldsymbol{x}_f) \, \mathrm{d}s_{\boldsymbol{x}} \right]}_{=0} n_i (\nabla a_i)_f$$

$$= |S_f| \, (a_i)_f n_i. \tag{3.19}$$

Putting this together yields

$$\int_{V_P} \nabla \cdot \boldsymbol{a} \, \mathrm{d}\boldsymbol{x} \approx \sum_f \underbrace{|S_f| \boldsymbol{n}}_{:=\boldsymbol{S}_f} \cdot \boldsymbol{a}_f = \sum_f \boldsymbol{S}_f \cdot \boldsymbol{a}_f. \tag{3.20}$$

The vector $\boldsymbol{S}_f$ in this equation is the surface vector, which is pointing into the direction of the outwards normal and has the same magnitude as the surface area. Alternative to the assumption of a linear variation over a cell the equations (3.14) and (3.19) could also be derived through numerical integration by using quadrature formulas with only one integration point (midpoint rule, see [34, §5.2]). Using the previous identities, every term of Equation (3.1) can be rewritten in its semi-discretized form. That means for the convective term

$$\int_{V_P} \nabla \cdot (\rho \boldsymbol{u} \phi) \, \mathrm{d}\boldsymbol{x} = \oint_{\partial V_P} (\rho \boldsymbol{u} \phi) \cdot \boldsymbol{n} \, \mathrm{d}s_{\boldsymbol{x}} = \sum_f \int_{S_f} (\rho \boldsymbol{u} \phi) \cdot \boldsymbol{n} \, \mathrm{d}s_{\boldsymbol{x}} \quad (3.21)$$

$$\approx \sum_f (\rho \boldsymbol{u} \phi)_f \boldsymbol{S}_f. \quad (3.22)$$

The diffusive term is transformed as

$$\int_{V_P} \nabla \cdot (\rho \Gamma_\phi \nabla \phi) \, \mathrm{d}\boldsymbol{x} = \oint_{\partial V_P} (\rho \Gamma_\phi \nabla \phi) \cdot \boldsymbol{n} \, \mathrm{d}s_{\boldsymbol{x}} \quad (3.23)$$

$$= \sum_f \int_{S_f} (\rho \Gamma_\phi \nabla \phi) \cdot \boldsymbol{n} \, \mathrm{d}s_{\boldsymbol{x}} \approx \sum_f (\rho \Gamma_\phi \nabla \phi)_f \cdot \boldsymbol{S}_f \quad (3.24)$$

and the source term and temporal term change similar to Equation (3.14):

$$\int_{V_P} S_\phi(\phi) \, \mathrm{d}\boldsymbol{x} \approx [S_\phi(\phi)]_P \, |V_P|, \quad (3.25)$$

$$\int_{V_P} \frac{\partial}{\partial t} (\rho \phi) \, \mathrm{d}\boldsymbol{x} \approx \left[\frac{\partial}{\partial t} (\rho \phi) \right]_P |V_P|. \quad (3.26)$$

Using all the results from (3.21), (3.23) and (3.25) in Equation (3.6) gives the semi-discretized transport equation

$$\int_t^{t+\Delta t} \left[\frac{\partial}{\partial t}(\rho\phi) \right]_P V_P \, \mathrm{d}t + \int_t^{t+\Delta t} \left[\sum_f (\rho \boldsymbol{u}\phi)_f \boldsymbol{S}_f \right] \mathrm{d}t$$

$$- \int_t^{t+\Delta t} \left[\sum_f (\rho \Gamma_\phi \nabla\phi)_f \boldsymbol{S}_f \right] \mathrm{d}t$$

$$= \int_t^{t+\Delta t} [S_\phi(\phi)]_P V_P \, \mathrm{d}t. \qquad (3.27)$$

Up to this point only the assumption of linear variation respectively the midpoint rule in space have been used. The truncated terms in Equations (3.7) and (3.8) are of second order in terms of spatial resolution and time step. Therefore the approximations leading to the semi-discretized Equation (3.27) are second order accurate. The next step is to derive a system of linear equations for the cell center values. This is sometimes referred to as flux linearization. For a further discretization in space the face center values should be expressed in terms of the cell center values and for the temporal term a suitable representation of the time derivative in terms of the discrete time steps is necessary. In the next sections a short overview of the discretization schemes used and their properties is given. This follows in most parts the descriptions in [16], [25] and [34]. In the used finite volume mesh any face belongs to a unique cell, its owner, given by the letter P. Therefore every inner face does also have a neighbor cell N and the respective normal is always directed from P to N.

After the second discretization step one will get a linear algebraic equation for every control volume. That means Equation (3.27) transforms to an equation of the form

$$a_P \phi_P + \sum_{N \in NB(P)} a_N \phi_N = b_P. \qquad (3.28)$$

Assembling all of the cell specific equations results in a system of linear equations. This system is usually solved by means of an iterative solution method because of the relatively high number of unknowns in CFD. The choice of iteration method depends on the structure of the coefficient matrix. An overview of different methods is given in [52]. If the matrix is symmetric one could use the (preconditioned) CG method. For general non-symmetric systems the BiCGSTAB ([56]) or a GAMG ([15], [13], [47]) method are possible choices. Therefore one should assure independently of the used solver that the iterative process will converge to the desired solution. A possibility to guarantee convergence for the iterative solution of linear problems is to keep the coefficient matrix (weakly) diagonal dominant (see [8, Satz 8.2.6 and Satz 8.2.9]). That means the coefficients have to satisfy

$$-\sum_{N \in NB(P)} a_N \leq a_P \text{ for all cells and}$$

$$-\sum_{N \in NB(P)} a_N < a_P \text{ at least for one cell.} \tag{3.29}$$

Whether the system is diagonally dominant or not is highly dependent on the final choice of discretization scheme respectively procedure. That means one should keep that in mind for the right choice of schemes. Examples for such discretization schemes are given in the following sections, starting with the diffusion term followed by the convection, source and temporal terms. Another important property of the resulting system of linear equations is that it should be consistent with the original differential equation and exhibit the same characteristics. Taking for example the steady pure diffusion equation without any source terms

$$-\nabla \cdot (\rho \Gamma_\phi \nabla \phi) = 0, \tag{3.30}$$

which can be deduced from Equation (3.1) by neglecting all the other terms, it can easily be seen that the solution is only defined up to a constant c. That means ϕ and $\phi + c$ are both solutions. In this case, the resulting linear equations should also have this property, therefore

$$a_P\phi_P + \sum_{N\in NB(P)} a_N\phi_N = 0 \text{ and}$$

$$a_P(\phi_P + c) + \sum_{N\in NB(P)} a_N(\phi_N + c) = 0,$$

should both hold true which gives

$$a_P + \sum_{N\in NB(P)} a_N = 0. \tag{3.31}$$

To guarantee consistency with (3.30) this should hold true in all the discretized equations. Expressing ϕ_P as

$$-\phi_P = \sum_{N\in NB(P)} \frac{a_N\phi_N}{a_P},$$

it can be interpreted as a weighted sum of its neighbors ([34, 8.2]). Therefore the value at P should be bounded by the values at its neighbors as long as no source terms are present.

One can assure this also with the opposite signs rule. In the above equations the coefficients of ϕ_P and ϕ_N are of opposite signs

$$a_P = - \sum_{N\in NB(P)} a_N.$$

This means an in- or decrease in ϕ_N will result in an in- or decrease of ϕ_P. This can be used as a sufficient condition to guarantee boundedness ([34, §8.2.2]). As long as the opposite signs rule is true the solution should remain bounded. If that is not the case it may get unbounded.

3.2.1 Diffusion Schemes

In the semi-discretized diffusion term given by Equation (3.23), the values of the (surface or normal) gradient at the faces of the mesh are

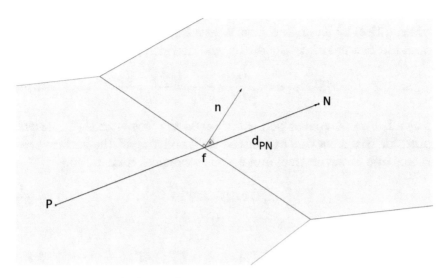

Figure 3.1: Non-Orthogonality

used. In order to derive a linear system of equations solely dependent on the cell centers an expression for those values in term of cell center values is needed. On orthogonal grids and with the assumption of a linear variation of ϕ it could be done in the following way

$$(\nabla \phi)_f \cdot \boldsymbol{S}_f = |\boldsymbol{S}_f| \frac{\phi_N - \phi_P}{\|\boldsymbol{x}_N - \boldsymbol{x}_P\|} = |\boldsymbol{S}_f| \frac{\phi_N - \phi_P}{d_{PN}}. \qquad (3.32)$$

The approach in Equation (3.32) uses the fact that on orthogonal grids the face normal $\boldsymbol{n}$ and the vector $\boldsymbol{d}_{PN}$ connecting the centroids of P and N are aligned. On arbitrary unstructured non-orthogonal grids this is generally no longer true (see Fig. 3.1). By computing the gradient according to (3.32) with the unit vector

$$\boldsymbol{e} = \frac{\boldsymbol{x}_N - \boldsymbol{x}_P}{\|\boldsymbol{x}_N - \boldsymbol{x}_P\|} = \frac{\boldsymbol{d}_{PN}}{d_{PN}}, \qquad (3.33)$$

which differs by an angle θ from $\boldsymbol{n}$ (see Figures 3.2, 3.3 and 3.4), one
would in fact calculate the directional derivative

$$(\nabla\phi)_f \cdot \boldsymbol{e} = \frac{\phi_N - \phi_P}{\|\boldsymbol{x}_N - \boldsymbol{x}_P\|} = \frac{\phi_N - \phi_P}{d_{PN}} \qquad (3.34)$$

which has a component perpendicular to the face normal. To accomplish for this error one can introduce a splitting of the surface area
vector into an orthogonal and a non-orthogonal contribution

$$\boldsymbol{S}_f = \boldsymbol{E}_f + \boldsymbol{T}_f \qquad (3.35)$$

and write the gradient as

$$(\nabla\phi)_f \cdot \boldsymbol{S}_f = (\nabla\phi)_f \cdot \boldsymbol{E}_f + (\nabla\phi) \cdot \boldsymbol{T}_f \qquad (3.36)$$

$$\approx \underbrace{E_f}_{=|\boldsymbol{E}_f|} \frac{\phi_N - \phi_P}{d_{PN}} + (\nabla\phi)_f \cdot \boldsymbol{T}_f. \qquad (3.37)$$

Therefore the diffusion term gets split into a linearized orthogonal
like part and a non-orthogonal part. This second term of (3.35) is
the so-called cross diffusion and leads to the idea of non-orthogonal
correction by using different options for this decomposition. These
can for example be the
Minimum Correction approach (Fig. 3.2) where

$$\boldsymbol{E}_f = (|S_f| \cos\theta)\boldsymbol{e} = (\boldsymbol{e} \cdot \boldsymbol{S}_f)\boldsymbol{e}, \qquad (3.38)$$

and

$$(\nabla\phi)_f \cdot \boldsymbol{T}_f = (\nabla\phi)_f \cdot (\boldsymbol{S}_f - \boldsymbol{E}_f) = (\nabla\phi)_f \cdot (\boldsymbol{n} - \cos\theta\boldsymbol{e})\,|S_f|, \qquad (3.39)$$

the **Orthogonal Correction approach** (Fig. 3.3) with

$$\boldsymbol{E}_f = |S_f|\,\boldsymbol{e} \qquad (3.40)$$

and

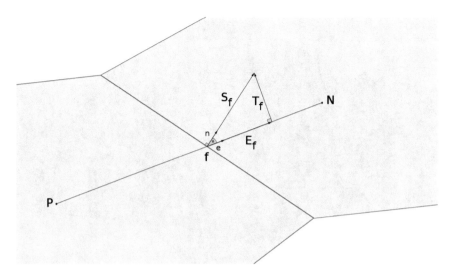

Figure 3.2: Minimum Correction

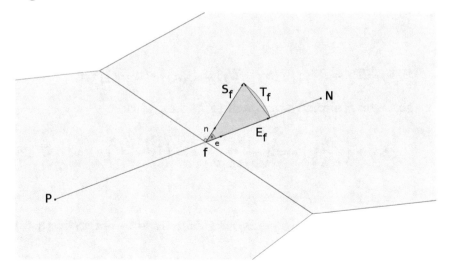

Figure 3.3: Orthogonal Correction

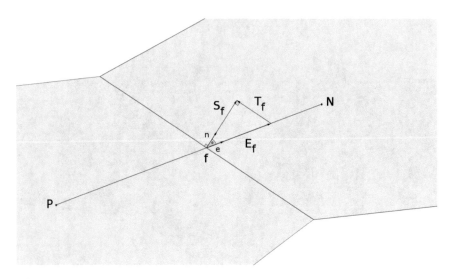

Figure 3.4: Over-Relaxed Correction

$$(\nabla\phi)_f \cdot \boldsymbol{T}_f = (\nabla\phi)_f \cdot (\boldsymbol{S}_f - \boldsymbol{E}_f) = (\nabla\phi)_f \cdot (\boldsymbol{n} - \boldsymbol{e}) \, |S_f| \,, \qquad (3.41)$$

or the **Over-Relaxed approach** (Fig. 3.4), where

$$\boldsymbol{E}_f = \left(\frac{|S_f|}{\cos\theta} \right) \boldsymbol{e} = \left(\frac{|S_f|^2}{S_f \cos\theta} \right) \boldsymbol{e} = \left(\frac{\boldsymbol{S}_f \cdot \boldsymbol{S}_f}{\boldsymbol{S}_f \boldsymbol{e}} \right) \boldsymbol{e} \qquad (3.42)$$

and

$$(\nabla\phi)_f \cdot \boldsymbol{T}_f = (\nabla\phi)_f \cdot (\boldsymbol{S}_f - \boldsymbol{E}_f) = (\nabla\phi)_f \cdot (\boldsymbol{n} - \frac{1}{\cos\theta}\boldsymbol{e}) \, |S_f| \,. \quad (3.43)$$

The evaluation of the cross-diffusion term needs again the value of the gradient at the cell faces. Those gradients could simply be neglected. That is mostly done in cases of small angles between $\boldsymbol{e}$ and $\boldsymbol{n}$ (low non-orthogonality) where the resulting error will be small. For higher non-orthogonality it can be computed by calculating the gradient at the cell centers using a mean value approximation with

$$(\nabla\phi)_P \approx \frac{1}{V_P} \sum_f \boldsymbol{S}_f \phi_f \qquad (3.44)$$

and interpolating those values to the faces. With a linear interpolation profile and a suitable geometric weighting factor $f_{\boldsymbol{x}}$ defined as the ratio of the distances from P to f and P to N with

$$f_{\boldsymbol{x}} := \frac{d_{Pf}}{d_{PN}}, \qquad (3.45)$$

this looks like

$$(\nabla\phi)_f = f_{\boldsymbol{x}}(\nabla\phi)_P + (1 - f_{\boldsymbol{x}})(\nabla\phi)_N \qquad (3.46)$$

This kind of gradient computation could also be used directly in (3.23) for the computation of the diffusion term, but that would include a significant higher number of computational nodes involved than the method described above. That is, because in Equation (3.44) the values of ϕ at all the faces are required, which again requires an interpolation using the values at all neighboring cell centers. To avoid this problem in the cross diffusion term the gradient is there computed in a deferred-correction approach, which means the gradient values from the previous iteration or time step are used. Following the tests in [25] the over-relaxed approach is found to be performing better than the others in terms of stability and accuracy. According to those results its convergence behavior is less oscillatory and it is able to converge at higher angles of non-orthogonality than the other approaches. Additionally the angle at which the non-orthogonal correction becomes necessary is higher. This approach is also followed in the OpenFOAM® implementation of the FVM ([44]). Besides the non-orthogonality there another general problem with the interpolation of cell values to the faces. For the midpoint integration to be accurate at the face it is essential that the correct value at the face midpoint is used. However, in a skewed mesh, the connection

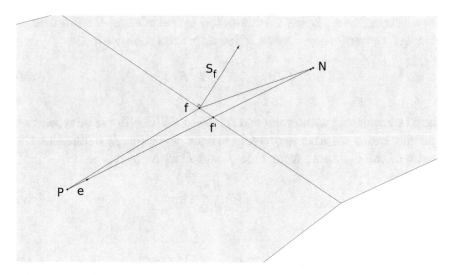

Figure 3.5: Skewness

between neighboring cell centers P and N does not cross the face exactly at the center. So the interpolation is done at another point $\boldsymbol{x}'_f$ instead (see Fig. 3.5). This can be corrected to some extend as long as the skewness is not too big by using an incomplete Taylor series expansion at $\boldsymbol{x}'_f$:

$$\phi_f \approx \phi'_f + (\nabla\phi)'_f \cdot \boldsymbol{d}'_f, \tag{3.47}$$

where $\boldsymbol{d}'_f$ is the vector connecting those two points. This is referred to as skewness correction.

3.2.2 Convection Schemes

In Equation (3.21) the convective term was transformed so that instead of the divergence of ϕ over the whole cell only the face values are needed. The (convective) mass flux through the face is defined as

$$F_f^{conv} := (\rho\boldsymbol{u})_f \cdot \boldsymbol{S}_f. \tag{3.48}$$

With Equation (3.21) the convective term can be written as

$$\int_{V_P} \nabla \cdot (\rho \boldsymbol{u} \phi) \, d\boldsymbol{x} = \sum_f F_f^{conv} \phi_f. \tag{3.49}$$

The calculation of the mass flux for those cases where the necessary values are either not known before or depending on the solution (e.g. in the momentum equation) will be explained later on depending on the specific situation. The remaining face value ϕ_f is computed by a so-called convection differencing scheme, which is used to express the face values in terms of cell values. One of the possibilities is using linear interpolation or the Central Differencing scheme (**CD**) like we did for the diffusion term. It is defined as

$$\phi_f = f_{\boldsymbol{x}} \phi_P + (1 - f_{\boldsymbol{x}}) \phi_N. \tag{3.50}$$

The factor $f_{\boldsymbol{x}}$ is the same as in Equation (3.45). This scheme represents a second order accurate calculation of the face center value ([34] and [16]). However, it has been observed that this scheme leads to unphysical oscillations in situations where convection is very important. In fact, the solution can become unbounded because of those under- and overshoots (see [16, section 4.7]). This problem can be overcome by the use of upwind differencing (**UD**), where the face value is determined by the direction of the flow/face flux. It is defined as

$$\phi_f = \begin{cases} \phi_P, & \text{if } F_f^{conv} \geq 0, \\ \phi_N, & \text{if } F_f^{conv} < 0. \end{cases} \tag{3.51}$$

The drawback of this scheme is that it is only first order accurate ([34], [16]) and tends to be very diffusive compared to **CD**. That can result in smoothing out local peaks or areas with high gradients. However, this two schemes are the most basic schemes and very easy to use in a computer code. The aim of overcoming the numerical shortcomings of the **CD** and **UD** schemes led to the construction of high order (**HO**) and high resolution (**HR**) schemes (see [34, §12]). Most of these schemes are of higher accuracy than **UD** while maintaining a

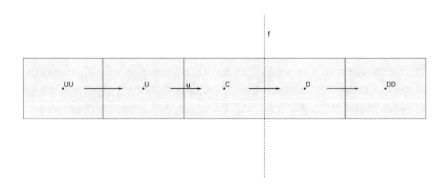

Figure 3.6: one dimensional domain used for the convection equation

stability similar to that of the upwind scheme. The cost of this is their increase in complexity compared to the basic schemes. To keep it simple the construction of **HO** and **HR** schemes is shown for the one dimensional pure convection equation without source terms

$$\frac{\partial \rho \phi}{\partial t} = -\frac{\partial \rho u \phi}{\partial x}. \qquad (3.52)$$

Details on the implementation of **HO** schemes on multidimensional nonuniform unstructured grids can be found in [34, §11.7-11.8, §12.6-12.8]. This equation is discretized on the domain depicted in Figure 3.6. The main cell is denoted by C and the velocity u is assumed to be positive. Therefore U, UU mean the upwind and far upwind nodes and D, DD are the downwind and far downwind nodes.

These cells are all uniform sized so that the faces are lying on the half distance between two cell centers. An example for a high order scheme is the so called second order upwind scheme (**SOU**, see[34]). It is constructed using an upwind based linear interpolation. The calculation of the face value is done by fitting a linear profile between the nodes U and C. This gives the formula

$$\phi(x) = \phi_C + \frac{\phi_C - \phi_U}{x_C - x_U}(x - x_C). \tag{3.53}$$

Using this equation for the uniform grid depicted in Fig. 3.6 gives the face value

$$\phi_f = \frac{3}{2}\phi_C - \frac{1}{2}\phi_U. \tag{3.54}$$

Computing ϕ_f with the **SOU** scheme is in fact not really an interpolation, it is more like an extrapolation because the face value is computed depending on the cell values of the upwind cells. The problem with this scheme is, that it is, although second order accurate, not unconditionally stable for general flow situations ([34]). The wide class of **HR** schemes is build with the aim of systematically getting this second order accuracy without introducing to much dispersion by limiting the dispersive portion of the scheme. **HR** schemes are often explained with the Normalized Variable Formulation (**NVF**, [28, 29, 30]) or the Total Variation Diminishing framework (**TVD**, see [53] or [34]). Here the **TVD** method is shortly explained. The Total Variation is defined as

$$TV(\phi) = \sum_i |\phi_{i+1} - \phi_i| \tag{3.55}$$

where the summation is done over the indices of all computational nodes (cells) in the domain. Any method is Total Variable Diminishing if

$$TV(\phi^{t+\Delta t}) \leq TV(\phi^t). \tag{3.56}$$

That means that $TV(\phi)$ does not increase with time. The face value ϕ_f could, interpolated with the **CD** scheme and written as the sum of an upwind interpolation and an additional term, be written as

$$\phi_f = \frac{1}{2}(\phi_D + \phi_C) = \phi_C + \frac{1}{2}(\phi_D - \phi_C) \tag{3.57}$$

The second term on the right hand side is some kind of anti-diffusive flux. That term is the part of the **CD** scheme which makes it second

order accurate. However, it is also responsible for the anti-diffusivity
which tends to produce oscillations. The aim in constructing high
resolution schemes that way is to add a part of this flux to the upwind
value which is small enough to avoid oscillations while preserving the
second order accuracy. That is done by introducing a flux limiter
$\psi(r)$. The face value is then written as

$$\phi_f = \phi_C + \frac{1}{2}\psi(r_f)(\phi_D - \phi_C) \tag{3.58}$$

with
$$r_f = \frac{\phi_C - \phi_U}{\phi_D - \phi_C}$$

as the ratio of two consecutive gradients. The limiter is assumed to
be non-negative. According to [34, 12.4] for a scheme to be **TVD**, it
is necessary that $\psi(r)$ fulfills the following inequality ([19],[53] and
[34])

$$0 \le \psi(r) - \frac{\psi(r)}{r}. \tag{3.59}$$

By assuming a value of 0 for negative values of r this holds if

$$\psi(r) \le 2 \text{ and } \psi(r) \le 2r. \tag{3.60}$$

This can be visually depicted in a $r - \psi$ or Sweby's diagram. Such
a diagram is shown in Figure 3.7 where the red area shows all the
values for which equations (3.60) hold true. Any flux limiter lying
in this region gives a **TVD** scheme. The limiter function for **UD** is
obviously $\psi(r) = 0$ and the limiter for **CD** can be seen from Equation
(3.57) as $\psi(r) = 1$. The **SOU** scheme can be rewritten as

$$\phi_f = \frac{3}{2}\phi_C - \frac{1}{2}\phi_U = \phi_C + \frac{1}{2}(\phi_C - \phi_U) \tag{3.61}$$

$$= \phi_C + \frac{1}{2}\frac{\phi_C - \phi_U}{\phi_D - \phi_C}(\phi_D - \phi_C) = \phi_C + \frac{1}{2}r_f(\phi_D - \phi_C) \tag{3.62}$$

which gives the limiter as $\psi(r) = r$. Drawing this in the diagram (Fig.
3.8) shows that the three schemes are not completely in the TVD-
region. Additionally we can see that the two second order schemes are

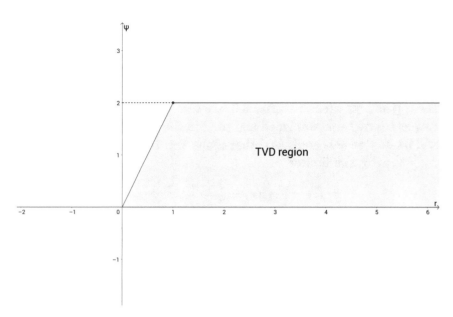

Figure 3.7: TVD region

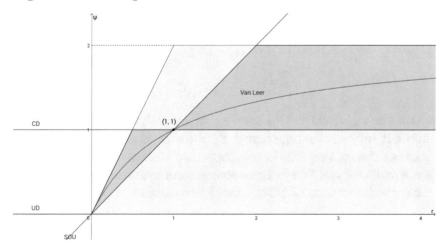

Figure 3.8: second order region

both going through the point $(1,1)$. In [57] it is shown that any second order scheme is in fact a weighted average between the **SOU** and **CD**. This means also that the limiter of every second order scheme has to go through this point $(1,1)$ and should lie in the area between them (the blue area in Figure 3.8, see also [34]). Many different **HR** schemes have been developed and used in the past years. One of them will be used in this study and that is the Van Leer scheme, introduced by [57], with the limiter

$$\psi(r) = \frac{r + |r|}{1 + |r|}.\tag{3.63}$$

The corresponding green line in Figure 3.8 is lying in the **TVD** region and in the second order area. That indicates that it is second order accurate and fulfills the **TVD** criteria from equations (3.56) and (3.59).

3.2.3 Source Term

The source term $S_\phi(\phi)$ is in general also depending on ϕ itself. It can be directly evaluated as in Equation (3.25) using the available values ϕ^* from previous iterations or time steps

$$\int_{V_P} S_\phi(\phi)\, \mathrm{d}\boldsymbol{x} = [S_\phi(\phi_P^*)]_P\, |V_P|$$

where the superscript * indicates this old value ([34, §14.1]). This does not introduce a big error if S_ϕ is either constant, very small or does not change very fast in comparison to all the other terms. If that is not true one can linearize the source term over a control volume by using the first terms of a Taylor series expansion in ϕ_P^* ([34])

$$S_\phi(\phi) \approx S_\phi(\phi_P^*) + \left(\frac{\partial S_\phi}{\partial \phi}\right)^* (\phi - \phi_P^*).\tag{3.64}$$

Using this relation in Equation (3.25) leads to the representation

$$S_\phi(\phi_P)\, |V_P| = \left(\frac{\partial S_\phi}{\partial \phi_P}\right)^* |V_P|\, \phi_P + \left(S_\phi(\phi_P^*) - \left(\frac{\partial S_\phi}{\partial \phi_P}\right)^* \phi_P^*\right) |V_P|\tag{3.65}$$

where the first term on the right hand side is treated implicitly while for the second, explicit part the old values are used.

3.2.4 Time Discretization

Before any integration in time the integral scalar transport equation on a control volume can be written in an abbreviated form as

$$\int_{V_P} \frac{\partial \rho \phi}{\partial t} \, d\boldsymbol{x} + \int_{V_P} \mathcal{L}(\phi) \, d\boldsymbol{x} = 0. \tag{3.66}$$

Here the operator $\mathcal{L}(\phi)$ includes all the spatial terms ([34, §14]). By applying a spatial discretization as described in the previous sections this changes to

$$\frac{\partial \rho_P \phi_P}{\partial t} V_P + L(\phi_P^t) = 0 \tag{3.67}$$

where $L(\phi_P^t)$ is the spatial discretization operator evaluated at a time t. This operator can also be expressed in its algebraic form as

$$L(\phi_P^t) = a_P \phi_P^t + \sum_{N \in NB(P)} a_N \phi_N^t - b_P^t. \tag{3.68}$$

The coefficients of this system of equations are in general dependent on the choice of spatial discretization schemes. The temporal grid is structured due to its one-dimensionality and very often uniform sized time-steps are used. Therefore it is an obvious choice to do the discretization of the temporal derivative with a finite difference approach. Many different schemes have been used for that and some of them will be presented here. The first explained schemes are the explicit and implicit Euler.

Using a time step Δt, one gets the explicit or forward Euler scheme by writing the value of an arbitrary function ϕ at the instant $t + \Delta t$ with a Taylor series approximation at t. That means

$$\phi(t + \Delta t) = \phi(t) + \frac{\partial \phi(t)}{\partial t} \Delta t + \frac{\partial^2 \phi(t)}{\partial t^2} \frac{\Delta t^2}{2} + \cdots, \tag{3.69}$$

which can be rewritten by neglecting all terms of order 2 or higher as

$$\frac{\partial \phi(t)}{\partial t} = \frac{\phi(t + \Delta t) - \phi(t)}{\Delta t} + \mathcal{O}(\Delta t) \qquad (3.70)$$

and is therefore first order accurate. Substituting that back into Equation (3.67) gives

$$\frac{(\rho_P \phi_P)^{t+\Delta t} - (\rho_P \phi_P)^t}{\Delta t} |V_P| + L(\phi_P^t) = 0 \qquad (3.71)$$

which can be written in algebraic form as

$$a_P^{t+\Delta t} \phi_P^{t+\Delta t} = b_P^t - \left((a_P^t + a_P) \phi^t + \sum_{N \in NB(P)} a_N \phi_N^t \right). \qquad (3.72)$$

This is an explicit time scheme and therefore computationally very cheap because all the new values can be computed from the existing values without the need to fully solve a linear system. The additional coefficients are

$$a_P^{t+\Delta t} = \frac{\rho_P^{t+\Delta t} |V_P|}{\Delta t}$$

$$a_P^t = -\frac{\rho_P^t |V_P|}{\Delta t}.$$

Although this scheme is very fast it is limited in terms of the maximum usable time step size. That is because in an explicit scheme the time step has to fulfill the so-called CFL criterion (originally reported in [10]). This criterion says that the discrete difference equation has to use at least all the information contained in the area of influence for the original differential equation in the same time interval. According to [34] the CFL criterion can be interpreted depending on the algebraic coefficients as the opposite signs rule extended to the temporal neighbors. In this case the main coefficient is $a_P^{t+\Delta t}$ whereas the temporal neighbor is $a_P + a_P^t$ and to assure boundedness the inequality

$$a_P^t + a_P \leq 0 \qquad (3.73)$$

should hold true. For a one dimensional pure convection problem spatially discretized with the upwind scheme this results in the very often used form

$$\frac{u_P \Delta t}{\Delta x_P} \leq 1, \tag{3.74}$$

see [34, p. 494ff]. The term on the left hand side is also called CFL or Courant number. This is a very strong limitation on the time step for the explicit Euler scheme especially for small cells. This scheme is for example used in the **MULES** limiter in the algorithm for the phase fraction field (see §3.4).

An easy way to overcome the problem with the CFL condition is the usage of the implicit or backward Euler scheme. For any function ϕ this is constructed by expressing the value at t with a Taylor series at $t + \Delta t$

$$\phi(t) = \phi(t + \Delta t) - \frac{\partial \phi(t + \Delta t)}{\partial t} \Delta t + \frac{\partial^2 \phi(t + \Delta t)}{\partial t^2} \frac{\Delta t^2}{2} + \cdots . \tag{3.75}$$

By ignoring the terms of order higher than Δt^2, the first derivative can be approximated as

$$\frac{\partial \phi(t + \Delta t)}{\partial t} = \frac{\phi(t + \Delta t) - \phi(t)}{\Delta t} + \mathcal{O}(\Delta t) \tag{3.76}$$

which results in a first order truncation error. By using Equation (3.76) in Equation (3.67) (evaluated at $t + \Delta t$) one gets

$$\frac{(\rho_P \phi_P)^{t + \Delta t} - (\rho_P \phi_P)^t}{\Delta t} |V_P| + L(\phi_P^{t + \Delta t}) = 0. \tag{3.77}$$

The corresponding algebraic form is

$$(a_P^{t + \Delta t} + a_P)\phi_P^{t + \Delta t} + \sum_{N \in NB(P)} a_N \phi_N^{t + \Delta t} = b_P^{t + \Delta t} + a_P^t \phi^t \tag{3.78}$$

with the additional coefficients

$$a_P^{t+\Delta t} = \frac{\rho_P^{t+\Delta t} |V_P|}{\Delta t}$$

$$a_P^t = -\frac{\rho_P^t |V_P|}{\Delta t}$$

This is a fully implicit scheme, therefore it is necessary to solve a full system of equations to march forward in time. However, the implicit Euler scheme is unconditionally stable independent of the time step size. This can be shown using the opposite signs rule analogue to the explicit scheme. The diagonal coefficient is here $a_P^{t+\Delta t} + a_P$ and the temporal neighbor is a_P^t. Therefore they are always of opposite signs ([34, §13.2.3]). Although there are larger time steps allowed with the backward Euler scheme it remains only first order accurate, but it is a very robust and simple scheme.

The wish for a scheme that is stable for larger time steps and second order accurate leads to the Crank-Nicolson scheme ([11]). Taking the Taylor series approximations for $\phi(t + \Delta t)$ and $\phi(t)$ at the instant $t + \frac{\Delta t}{2}$

$$\phi(t + \Delta t) = \phi(t + \frac{\Delta t}{2}) + \frac{\partial \phi(t + \frac{\Delta t}{2})}{\partial t} \frac{\Delta t}{2} + \frac{\partial^2 \phi(t + \frac{\Delta t}{2})}{\partial t^2} \frac{\Delta t^2}{8} + \cdots ,$$
$$(3.79)$$

$$\phi(t) = \phi(t + \frac{\Delta t}{2}) - \frac{\partial \phi(t + \frac{\Delta t}{2})}{\partial t} \frac{\Delta t}{2} + \frac{\partial^2 \phi(t + \frac{\Delta t}{2})}{\partial t^2} \frac{\Delta t^2}{8} + \cdots ,$$
$$(3.80)$$

and subtracting (3.80) from (3.79) one obtains an expression for the first derivative at $t + \frac{\Delta t}{2}$ that is second order accurate

$$\frac{\partial \phi(t + \frac{\Delta t}{2})}{\partial t} = \frac{\phi(t + \Delta t) - \phi(t)}{\Delta t} + \mathcal{O}(\Delta t^2). \qquad (3.81)$$

Inserting this in Equation (3.67) (evaluated at $t + \frac{\Delta t}{2}$) gives

$$\frac{(\rho_P \phi_P)^{t+\Delta t} - (\rho_P \phi_P)^t}{\Delta t} |V_P| + L(\phi_P^{t+\frac{\Delta t}{2}}) = 0, \qquad (3.82)$$

which results in an algebraic system of the form

$$
a_P^{t+\Delta t}\phi_P^{t+\Delta t} = b_P^{t+\frac{\Delta t}{2}} - \left(a_P^{t+\frac{\Delta t}{2}}\phi_P^{t+\frac{\Delta t}{2}} + \sum_{N\in NB(P)} a_N\phi_N^{t+\frac{\Delta t}{2}} \right) - a_P^t\phi^t,
$$

(3.83)

where the temporal coefficients are given by ([34, §13.2.4])

$$
a_P^{t+\Delta t} = \frac{\rho_P^{t+\Delta t}|V_P|}{\Delta t},
$$
$$
a_P^t = -\frac{\rho_P^t|V_P|}{\Delta t}.
$$

By using a midpoint approximation for

$$
\phi^{t+\frac{\Delta t}{2}} \approx \frac{\phi^{t+\Delta t} + \phi^t}{2}
$$

and

$$
b^{t+\frac{\Delta t}{2}} \approx \frac{b^{t+\Delta t} + b^t}{2}
$$

the above equation can be written as

$$
a_P^{t+\Delta t}\phi_P^{t+\Delta t} + \frac{1}{2}\left(a_P\phi_P^{t+\Delta t} + \sum_{N\in NB(P)} a_N\phi_N^{t+\Delta t} - b_P^{t+\Delta t} \right)
$$
$$
= \frac{1}{2}\left(b_P^t - a_P\phi_P^t + \sum_{N\in NB(P)} a_N\phi_N^t \right) - a_P^t\phi^t
$$
$$
= \frac{1}{2}\left(b_P^t - (a_P + 2a_P^t)\phi_P^t + \sum_{N\in NB(P)} a_N\phi_N^t \right)
$$

(3.84)

or in the general form as

$$
\frac{(\rho_P\phi_P)^{t+\Delta t} - (\rho_P\phi_P)^t}{\Delta t}|V_P| + \frac{1}{2}\left[L(\phi_P^{t+\Delta t}) + L(\phi_P^t) \right] = 0.
$$

(3.85)

In (3.84) the temporal neighbor can be identified as $a_P + 2a_P^t$ and should remain negative to guarantee stability. For a one dimensional convection case where the spatial terms are discretized in the same way as for Equation (3.74) this gives the inequality

$$\frac{u_P \Delta t}{\Delta x_P} \leq 2 \qquad (3.86)$$

(see [34, §13.2.4]). Therefore this scheme allows twice the time step size of the explicit Euler scheme and is second order accurate, but according to Equation (3.84) it has an implicit part like the backward Euler scheme. Therefore a system of linear equations has to be solved at every time step. However, it is possible to avoid this by using the right implementation. The Crank-Nicolson scheme can be implemented as a two step procedure starting with the implicit Euler followed by the explicit Euler ([34, §13.2.5]). By using an intermediate time step $\frac{\Delta t}{2}$ with Equation (3.77) one gets

$$\frac{(\rho_P \phi_P)^{t+\frac{\Delta t}{2}} - (\rho_P \phi_P)^t}{\frac{\Delta t}{2}} |V_P| = -L(\phi_P^{t+\frac{\Delta t}{2}}). \qquad (3.87)$$

doing the same with Equation (3.71) starting from this intermediate time-step gives

$$\frac{(\rho_P \phi_P)^{t+\Delta t} - (\rho_P \phi_P)^{t+\frac{\Delta t}{2}}}{\frac{\Delta t}{2}} |V_P| = -L(\phi_P^{t+\frac{\Delta t}{2}}). \qquad (3.88)$$

Adding up these equations leads to

$$\frac{(\rho_P \phi_P)^{t+\Delta t} - (\rho_P \phi_P)^{t+\frac{\Delta t}{2}}}{\frac{\Delta t}{2}} |V_P| + \frac{(\rho_P \phi_P)^{t+\frac{\Delta t}{2}} - (\rho_P \phi_P)^t}{\frac{\Delta t}{2}} |V_P|$$

$$\qquad (3.89)$$

$$= -2L(\phi_P^{t+\frac{\Delta t}{2}}) \qquad (3.90)$$

or

$$\frac{(\rho_P \phi_P)^{t+\Delta t} - (\rho_P \phi_P)^t}{\Delta t} |V_P| = -L(\phi_P^{t+\frac{\Delta t}{2}}),$$

which is equal to the Crank-Nicolson Scheme in (3.82). In a practical implementation this two step procedure is done by combining equations (3.87) and (3.88) in the following way

$$\frac{(\rho_P\phi_P)^{t+\Delta t} - (\rho_P\phi_P)^{t+\frac{\Delta t}{2}}}{\frac{\Delta t}{2}} |V_P| = \frac{(\rho_P\phi_P)^{t+\frac{\Delta t}{2}} - (\rho_P\phi_P)^{t}}{\frac{\Delta t}{2}} |V_P|$$

(3.91)

which can be simplified to

$$(\rho_P\phi_P)^{t+\Delta t} = 2(\rho_P\phi_P)^{t+\frac{\Delta t}{2}} - (\rho_P\phi_P)^{t}.$$

(3.92)

Implemented in this way the Crank-Nicolson scheme is fully explicit, but the values of the last two time-steps are always needed. Comparing equations (3.71), (3.77) and (3.82) one can describe all three schemes in a unified approach with

$$\frac{(\rho_P\phi_P)^{t+\Delta t} - (\rho_P\phi_P)^{t}}{\Delta t} V_P + \alpha L(\phi_P^{t+\Delta t}) + (1-\alpha)L(\phi_P^t) = 0.$$

(3.93)

where $\alpha \in \mathbb{R}$ is some kind of blending parameter and normally between 0 and 1. By using the values 0, $\frac{1}{2}$ or 1 one gets the explicit Euler, the Crank-Nicolson or the implicit Euler scheme as special cases of this general approach. As mentioned in [25], the Crank-Nicolson scheme does not guarantee boundedness of the solution. In OpenFOAM® the Crank-Nicolson scheme is implemented as a blending between the pure Crank-Nicolson and the implicit Euler scheme. Therefore it is possible to stabilize the solution by decreasing the usage of Crank-Nicolson while maintaining some portion of the second-order accuracy ([36]). An additional option in OpenFOAM® is the possible usage of adaptive time steps, where the size of the next time interval is defined by the current velocity (flux) field and a desired maximal CFL number ([36], [12]).This is of great help if the maximal usable time step size is not known before and for maintaining a small CFL number for the explicit solution of the phase fraction Equation (§3.4). The usage of the Euler schemes with non uniform time steps is straight forward because both do only need the information of one time step. The

Crank-Nicolson schemes needs to be adapted for that. Details on the Crank-Nicolson scheme with non-uniform time steps can be found in [34, §13.4.1].

3.3 Discretization of the Navier-Stokes equations

The next step is the application of the general discretization techniques from the previous chapter to the incompressible Navier-Stokes equations, given by Equations (N.-S.). Using the described interpolation techniques directly can result in decoupling of the pressure and velocity fields (see [34, §15.5.1]). One possibility to overcome this problem is the usage of staggered grids where the pressure and all the components of the velocity field are calculated and stored on different (shifted) grid systems. Because of this extra grid systems the needed computational storage increases significantly and the use of general unstructured meshes leads to a higher number of equations to be solved. However, with the Rhie-Chow interpolation technique ([43]), it is possible to avoid the necessity of a staggered variable arrangement. This interpolation addresses and removes the decoupling problem while maintaining the advantages of a collocated grid ([34, §15.5]). The flow equations are then discretized using a segregated approach. The methods used in this work are from the **SIMPLE** (Semi Implicit Method for Pressure Linked Equations) family of algorithms. This class of algorithms is based on the original **SIMPLE** algorithm ([39], [40], [41]) or a bit more precisely: the different formulations are all sharing the same basic derivations which are presented here according to [34].

Basically the discretization starts by using Gauss Theorem to rewrite the mass conservation equation with the face volume flux F_f and the face mass flux $\dot{m}_f$ as

$$\int_{V_P} \nabla \cdot (\rho \boldsymbol{u}) \ \mathrm{d}\boldsymbol{x} = \sum_{f \in F(P)} \int_{S_f} (\rho \boldsymbol{u}) \cdot \boldsymbol{n} \ \mathrm{d}s_{\boldsymbol{x}}$$

$$\approx \sum_{f \in F(P)} \rho_f \underbrace{\boldsymbol{u}_f \cdot \boldsymbol{S}_f}_{:=F_f} = \sum_{f \in F(P)} \underbrace{\rho_f F_f}_{:=\dot{m}_f}$$

$$= \sum_{f \in F(P)} \dot{m}_f = 0, \tag{3.94}$$

where $F(P)$ means all the faces belonging to cell P. The momentum equations are taken in the form

$$\int_{V_P} \frac{\partial \rho \boldsymbol{u}}{\partial t} \ \mathrm{d}\boldsymbol{x} + \int_{V_P} \nabla \cdot \{\rho \boldsymbol{u}\boldsymbol{u}\} \ \mathrm{d}\boldsymbol{x} = -\underline{\int_{V_P} \nabla p \ \mathrm{d}\boldsymbol{x}} \tag{3.95}$$

$$+ \int_{V_P} \nabla \cdot \{\mu \nabla \boldsymbol{u}\} \ \mathrm{d}\boldsymbol{x} + \underline{\int_{V_P} \nabla \cdot \{\mu(\nabla \boldsymbol{u})^T\} \ \mathrm{d}\boldsymbol{x}} + \underline{\int_{V_P} \boldsymbol{f} \ \mathrm{d}\boldsymbol{x}}. \tag{3.96}$$

In the discretization process, the underlined terms are evaluated explicitly using old values of the required fields ([34, §15.5.1]) and therefore included in the source term. This can be done analogous to the sections 3.2.1 and 3.2.3 by either transforming them into surface integrals or approximating them by using the cell center values. This results in a system of linear equations for the velocity components for every cell P

$$a_P^{\boldsymbol{u}} \boldsymbol{u}_P + \sum_{N \in NB(P)} a_N^{\boldsymbol{u}} \boldsymbol{u}_N = \boldsymbol{b}_P^{\boldsymbol{u}}. \tag{3.97}$$

Taking the (semi-discretized) pressure term out of the source term

$$\boldsymbol{b}_P^{\boldsymbol{u}} = -|V_P| \, (\nabla p)_P + \hat{\boldsymbol{b}}_P^{\boldsymbol{u}}$$

and dividing by $a_P^{\boldsymbol{u}}$ gives

$$\boldsymbol{u}_P + \sum_{N \in NB(P)} \frac{a_N^{\boldsymbol{u}}}{a_P^{\boldsymbol{u}}} \boldsymbol{u}_N = -\frac{|V_P|}{a_P^{\boldsymbol{u}}} (\nabla p)_P + \frac{\hat{\boldsymbol{b}}_P^{\boldsymbol{u}}}{a_P^{\boldsymbol{u}}}. \tag{3.98}$$

This can be reformulated as

$$\boldsymbol{u}_P + \boldsymbol{H}_p[\boldsymbol{u}] = -\boldsymbol{D}_P^{\boldsymbol{u}}(\nabla p)_P + \boldsymbol{B}_P^{\boldsymbol{u}} \qquad (3.99)$$

with the vector operators

$$\boldsymbol{H}_p[\boldsymbol{u}] = \sum_{N \in NB(P)} \frac{a_N^{\boldsymbol{u}}}{a_P^{\boldsymbol{u}}} \boldsymbol{u}_N,$$

$$\boldsymbol{D}_P^{\boldsymbol{u}} = \frac{|V_P|}{a_P^{\boldsymbol{u}}} \text{ and}$$

$$\boldsymbol{B}_P^{\boldsymbol{u}} = \frac{\hat{\boldsymbol{b}}_P^{\boldsymbol{u}}}{a_P^{\boldsymbol{u}}}.$$

3.3.1 The Rhie-Chow Interpolation

The Rhie-Chow interpolation is used to get the face velocity values from the cell center values. This is necessary to avoid the possible decoupling of the pressure and velocity fields by using a simple linear interpolation. This decoupling can be easily demonstrated in 1D where so-called checkerboard fields (fields with alternating values from cell to cell) are misinterpreted as uniform fields by the **SIMPLE** algorithm (see [34, section 15.2.3]). Assuming a collocated grid arrangement on a 1D mesh with uniform cells of length Δx as in Figure 3.9 and specifying f' as the upwind face, the pressure term for the central cell V_C can be transformed to a semi-discretized form

$$\int_{V_C} \frac{\partial p}{\partial x} \, \mathrm{d}x \approx \int_{S_f} p \, \mathrm{d}s_x - \int_{S_{f'}} p \, \mathrm{d}s_x.$$

Since $|S_{f'}| = |S_f| = \frac{|V_C|}{\Delta x}$ this can be further simplified by approximation of the surface integrals with the midpoint rule to

$$\int_{V_C} \frac{\partial p}{\partial x} \, \mathrm{d}x \approx \frac{|V_C|}{\Delta x} \left[p_f - p_{f'} \right].$$

Computing the face values directly with linear interpolation gives

$$\int_{V_C} \frac{\partial p}{\partial x} \, \mathrm{d}x \approx \frac{|V_C|}{\Delta x} \left[\frac{1}{2}(p_D + p_C) - \frac{1}{2}(p_C + p_U) \right] = |V_C| \frac{p_D - p_U}{2\Delta x}$$

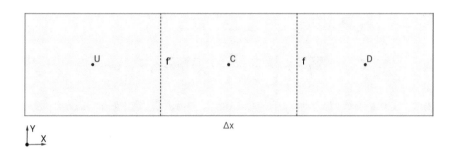

Figure 3.9: one dimensional domain used for the discretization of the pressure term

(see [34, §15.2.3]). Thus the equation for the central cell C does only include the pressure values of the neighboring cells. A similar situation can be derived for the velocity by using the continuity equation. In both cases only alternating values are linked to each other. If the fields have alternating values (zigzag or checkerboard fields) this would then be mistakenly seen as a uniform field by the numerical method. That is when the Rhie-Chow interpolation is needed.

Face values computed by linear interpolation will be expressed with an over-bar in the following derivations. That means for any function ϕ, a cell face f between cells P and N and the geometric factor f_x from Equation (3.45) the face value is evaluated as

$$\overline{\phi_f} = f_x \phi_P + (1 - f_x)\phi_N. \qquad (3.100)$$

The Rhie Chow interpolation was originally reported in [43] and will be presented here according to the derivations in [34]. In this interpolation technique an additional dissipation term containing two variants of the cell face pressure gradient is added to the linear interpolated velocity values. The face velocity is then expressed as

$$\boldsymbol{u}_f = \overline{\boldsymbol{u}_f} - \overline{\boldsymbol{D}_f^{\boldsymbol{u}}} \left(\nabla p_f - \overline{\nabla p_f} \right). \qquad (3.101)$$

The cell face pressure gradient ∇p_f is computed according to the gradient computation in section 3.2.1 as

$$\nabla p_f = \overline{\nabla p_f} + \left[\frac{p_N - p_P}{d_{PN}} - (\overline{\nabla p_f} \cdot e_{PN}) \right] e_{PN}.$$

That means that the pressure gradient in the PN direction is computed using the values of the adjacent cells

$$\nabla p_f \cdot e_{PN} = \overline{\nabla p_f} \cdot e_{PN} + \left[\frac{p_N - p_P}{d_{PN}} - (\overline{\nabla p_f} \cdot e_{PN}) \right] e_{PN} \cdot e_{PN}$$

$$= \frac{p_N - p_P}{d_{PN}}$$

and therefore the face velocities are also directly related to the adjacent pressure values. This makes checkerboard fields inadmissible and therefore removes the possibility of decoupling. Furthermore this makes the application of the **SIMPLE** algorithm on unstructured collocated grids possible.

3.3.2 The SIMPLE algorithm

The original **SIMPLE** algorithm ([39], [40], [41]) was first developed for a staggered grid arrangement. With the Rhie-Chow interpolation (see 3.3.1) it was possible to adapt the algorithm for collocated grids. The respective formulation on such grids is described following [34]. The algorithm starts with either guessed or old values $(\boldsymbol{u}^n, \dot{m}^n, p^n)$, where $\dot{m}$ is the mass flow rate at the boundaries of the cells. Equation (3.99) is then first solved for an intermediate velocity field $\boldsymbol{u}^*$

$$\boldsymbol{u}_P^* + \boldsymbol{H}_p[\boldsymbol{u}^*] = -\boldsymbol{D}_P^{\boldsymbol{u}}(\nabla p^n)_P + \boldsymbol{B}_P^{\boldsymbol{u}} \qquad (3.102)$$

using the old pressure field. This intermediate solution obeys momentum conservation but not necessarily mass conservation. Therefore some correction fields $(\boldsymbol{u}', \dot{m}', p')$ are needed to get both momentum and mass conservation. They are related to the desired solution according to

$$\boldsymbol{u} = \boldsymbol{u}^* + \boldsymbol{u}',$$
$$p = p^n + p',$$
$$\dot{m} = \dot{m}^* + \dot{m}'.$$

This gives

$$\sum_{f \in F(P)} \dot{m}'_f = - \sum_{f \in F(P)} \dot{m}^*_f \tag{3.103}$$

for the continuity equation, where $\dot{m}^*_f$ is computed as

$$\dot{m}^*_f = \rho_f \boldsymbol{u}^*_f \cdot \boldsymbol{S}_f. \tag{3.104}$$

The estimated face velocity $\boldsymbol{u}^*_f$ is computed with the Rhie-Chow interpolation as

$$\boldsymbol{u}^*_f = \overline{\boldsymbol{u}^*_f} - \overline{\boldsymbol{D}^{\boldsymbol{u}}_f} \left(\nabla p^n_f - \overline{\nabla p^n_f} \right). \tag{3.105}$$

The difference between equations (3.99) and (3.102) gives

$$\boldsymbol{u}'_P + \boldsymbol{H}_P[\boldsymbol{u}'] = -\boldsymbol{D}^{\boldsymbol{u}}_P (\nabla p')_P \tag{3.106}$$

and the correction of the mass flow rate can be written as

$$\dot{m}'_f = \rho_f \boldsymbol{u}'_f \cdot \boldsymbol{S}_f. \tag{3.107}$$

The correction $\boldsymbol{u}'_f$ of the cell face velocity value is computed by subtracting Equation (3.105) from Equation (3.101). This results in

$$\boldsymbol{u}'_f = \overline{\boldsymbol{u}'_f} - \overline{\boldsymbol{D}^{\boldsymbol{u}}_f} \left(\nabla p'_f - \overline{\nabla p'_f} \right). \tag{3.108}$$

Using Equations (3.107) and (3.108) in Equation (3.103) leads to

$$\sum_{f \in F(P)} (\rho_f \overline{\boldsymbol{u}'_f} \cdot \boldsymbol{S}_f) + \sum_{f \in F(P)} (\rho_f \overline{\boldsymbol{D}^{\boldsymbol{u}}_f \nabla p'_f} \cdot \boldsymbol{S}_f)$$
$$- \sum_{f \in F(P)} (\rho_f \overline{\boldsymbol{D}^{\boldsymbol{u}}_f} \nabla p'_f \cdot \boldsymbol{S}_f) = - \sum_{f \in F(P)} \dot{m}^*_f. \tag{3.109}$$

Taking the respective neighboring cell N corresponding to a face f, an equation for the mass flow rate correction can be constructed analogous to Equation (3.106)

$$\boldsymbol{u}_N' + \boldsymbol{H}_N[\boldsymbol{u}'] = -\boldsymbol{D}_N^{\boldsymbol{u}}(\nabla p')_N. \qquad (3.110)$$

Linear interpolation between Equation (3.106) and Equation (3.110) gives

$$\overline{\boldsymbol{u}_f'} + \overline{\boldsymbol{H}_f[\boldsymbol{u}']} = -\overline{\boldsymbol{D}_f^{\boldsymbol{u}}(\nabla p')_f}, \qquad (3.111)$$

which can be rewritten as

$$\overline{\boldsymbol{u}_f'} + \overline{\boldsymbol{D}_f^{\boldsymbol{u}}(\nabla p')_f} = -\overline{\boldsymbol{H}_f[\boldsymbol{u}']}. \qquad (3.112)$$

Inserting this in Equation (3.109) yields the pressure correction equation

$$\sum_{f\in F(P)} (-\rho_f \overline{\boldsymbol{D}_f^{\boldsymbol{u}}} \nabla p_f' \cdot \boldsymbol{S}_f) =$$

$$- \sum_{f\in F(P)} \dot{m}_f^* + \sum_{f\in F(P)} (\rho_f \underline{\overline{\boldsymbol{H}_f[\boldsymbol{u}']}} \cdot \boldsymbol{S}_f). \qquad (3.113)$$

The underlined term can not be directly used because it requires the unknown correction field $\boldsymbol{u}'$. Its particular treatment results in different variants of the **SIMPLE** algorithm [34, S 15.7]. In the original formulation of the algorithm this term is neglected. That does not change the final solution since it is a correction term in an iterative process which becomes zero for the converged solution. However, it does influence the convergence behavior and results in larger pressure corrections. Therefore often explicit under-relaxation is used for the pressure correction values. The treatment of the remaining terms in Equation (3.113) can be done with the methods from the previous section. The resulting coefficients for the linear equations are constructed as per the diffusion term discretization. That means the term on the left side is rewritten according to

$$(\overline{\boldsymbol{D_f^u}}\nabla p_f') \cdot \boldsymbol{S}_f = (\nabla p_f' \overline{\boldsymbol{D_f^u}}^T) \cdot \boldsymbol{S}_f \tag{3.114}$$

$$= \nabla p_f' \cdot (\overline{\boldsymbol{D_f^u}}^T \cdot \boldsymbol{S}_f) \tag{3.115}$$

$$= \nabla p_f' \cdot \boldsymbol{S}_f' \tag{3.116}$$

where

$$\boldsymbol{S}_f' = \overline{\boldsymbol{D_f^u}}^T \cdot \boldsymbol{S}_f = \begin{pmatrix} \overline{D_f^{u_1}} & 0 & 0 \\ 0 & \overline{D_f^{u_2}} & 0 \\ 0 & 0 & \overline{D_f^{u_3}} \end{pmatrix} \begin{pmatrix} S_f^x \\ S_f^y \\ S_f^z \end{pmatrix} = \begin{pmatrix} \overline{D_f^{u_1}} S_f^x \\ \overline{D_f^{u_2}} S_f^y \\ \overline{D_f^{u_3}} S_f^z \end{pmatrix}. \tag{3.117}$$

By using the modified surface vector $\boldsymbol{S}_f'$ this becomes a diffusion term as in Section 3.2.1 (with $\phi = p$) and can be discretized accordingly with one of the choices for non-orthogonality correction (Equations (3.39), (3.41) and (3.43)). Doing so results in an algebraic equation

$$a_P^{p'} p_P' + \sum_{N \in NB(P)} a_N^{p'} p_N' = \boldsymbol{b}_P^{p'} \tag{3.118}$$

with the term on the right side given as

$$\boldsymbol{b}_P^{p'} = -\sum_{f \in F(P)} \dot{m}_f^* + \underline{\sum_{f \in F(P)} (\rho_f \overline{\boldsymbol{H}}_f[\boldsymbol{u}'] \cdot \boldsymbol{S}_f)} \tag{3.119}$$

and the mass flow rate $\dot{m}_f^*$ computed with the Rhie-Chow interpolation as in Equation (3.105). As mentioned above, due to the neglection of the underlined term in the **SIMPLE** algorithm, under-relaxation of the pressure correction field can improve the convergence of the algorithm ([34, §15.5.2]). Using an under-relaxation factor λ^P, the following corrected values are obtained

$$\boldsymbol{u}_P^{**} = \boldsymbol{u}_P^* + \boldsymbol{u}_P' \qquad \text{with } \boldsymbol{u}_P' = -\boldsymbol{D}_P^u \nabla p_P' \tag{3.120}$$

$$\dot{m}_f^{**} = \dot{m}_f^* + \dot{m}_f' \qquad \text{with } \dot{m}_f' = -\rho_f \boldsymbol{D}_f^u \nabla p_f' \cdot \boldsymbol{S}_f \tag{3.121}$$

$$p_P^* = p_P^n + \lambda^P p_P' \tag{3.122}$$

Putting these steps together one step of the collocated **SIMPLE** algorithm can be summarized in the following way:

1. Start with a given solution $(\boldsymbol{u}^n, \dot{m}^n, p^n)$ at iteration n (for example time t) as initial guess.

2. Solve the momentum equation (Eq. (3.102)) to get a intermediate momentum conserving velocity field $\boldsymbol{u}^*$.

3. Use the obtained velocity field to compute an updated mass flow rate $\dot{m}^*$ with the Rhie-Chow interpolation (Equations (3.104) and (3.105)).

4. Construct the pressure correction equation (Eq. (3.113)) with the updated mass flow rate and solve it to get a pressure correction field p'.

5. Update the velocity and pressure fields using this pressure correction with Equation (3.120) to get the mass conserving fields $(\boldsymbol{u}^{**}, \dot{m}^{**}, p^*)$.

6. Take the obtained solution as a new initial guess and repeat from step 2 until convergence is achieved.

7. Use the converged solution as the exact solution at iteration $n+1$ (or time $t + \Delta t$) and proceed to the next iteration.

8. Repeat from step 1 until the last iteration or time step is reached.

3.3.3 The PISO algorithm

Starting with the **SIMPLE** algorithm, a whole family of similar algorithms have been developed (see [34, §15.7] for an overview of the algorithms and their differences). They are mostly differing in the treatment of the velocity correction term (the underlined term in Eq. (3.113)). The **PISO** (Pressure-Implicit Split Operator) algorithm is an often encountered member of this family. It is composed of two or more different steps. Starting from an initial guess first a full **SIMPLE** iteration is done to get the continuity satisfying fields $(\boldsymbol{u}^{**}, \dot{m}^{**}, p^*)$ by neglecting $\boldsymbol{H}_P[\boldsymbol{u}']$. In a second step this fields are used to reconstruct the momentum equation, but this time using the updated fields to compute the respective coefficients/operators $\boldsymbol{H}_p^{**}$, $(\boldsymbol{D}_P^{\boldsymbol{u}})^{**}$ and $(\boldsymbol{B}_P^{\boldsymbol{u}})^{**}$. This new momentum equation is then solved explicitly for every cell to derive a new velocity field $\boldsymbol{u}^{***}$

$$\boldsymbol{u}_P^{***} = -\boldsymbol{H}_p^{**}[\boldsymbol{u}^{**}] - (\boldsymbol{D}_P^{\boldsymbol{u}})^{**}(\nabla p^*)_P + (\boldsymbol{B}_P^{\boldsymbol{u}})^{**}. \tag{3.123}$$

This field is further used to compute a new mass flow rate $\dot{m}^{***}$ with the Rhie-Chow interpolation (analogous to Equation (3.104)). With that some part of the neglected $\boldsymbol{H}_P[\boldsymbol{u}']$ term is recovered by writing the remaining velocity correction as

$$\begin{aligned}
\boldsymbol{u}_P^{****} &= \boldsymbol{u}_P^{***} + \boldsymbol{u}_P'' \\
&= -\boldsymbol{H}_p^{**}[\boldsymbol{u}^{**}] - (\boldsymbol{D}_P^{\boldsymbol{u}})^{**}(\nabla p^*)_P + (\boldsymbol{B}_P^{\boldsymbol{u}})^{**} + \boldsymbol{u}_P'' \\
&= -\boldsymbol{H}_p^{**}[\boldsymbol{u}^* + \boldsymbol{u}'] - (\boldsymbol{D}_P^{\boldsymbol{u}})^{**}(\nabla p^*)_P + (\boldsymbol{B}_P^{\boldsymbol{u}})^{**} + \boldsymbol{u}_P'' \\
&= -\boldsymbol{H}_p^{**}[\boldsymbol{u}^*] - \boldsymbol{H}_p^{**}[\boldsymbol{u}'] - (\boldsymbol{D}_P^{\boldsymbol{u}})^{**}(\nabla p^*)_P + (\boldsymbol{B}_P^{\boldsymbol{u}})^{**} + \boldsymbol{u}_P'' \\
&= \underbrace{-\boldsymbol{H}_p^{**}[\boldsymbol{u}^*] - (\boldsymbol{D}_P^{\boldsymbol{u}})^{**}(\nabla p^*)_P + (\boldsymbol{B}_P^{\boldsymbol{u}})^{**}}_{\approx \boldsymbol{u}^{**}} - \boldsymbol{H}_p^{**}[-\boldsymbol{D}_P^{\boldsymbol{u}}\nabla p_P'] + \boldsymbol{u}_P''.
\end{aligned}$$

The first 3 terms on the right hand resemble an explicit evaluation of the momentum equation with updated coefficients and pressure values, but the old velocity values. Therefore it is an approximation of $\boldsymbol{u}^{**}$. This gives the new velocity correction as

$$\boldsymbol{u}_P^{****} \approx \boldsymbol{u}^{**} - \underline{\boldsymbol{H}_p^{**}[-\boldsymbol{D}_P^{\boldsymbol{u}}\nabla p_P']} + \underline{\boldsymbol{u}_P''}, \tag{3.124}$$

where the underlined terms represent the recovered part of the originally neglected term $\boldsymbol{H}_P[\boldsymbol{u}']$ ([34, §15.7.3]). The velocity correction of the second step, $\boldsymbol{u}_P''$, satisfies

$$\boldsymbol{u}_P'' = -\underline{\boldsymbol{H}_p^{**}[\boldsymbol{u}'']} - (\boldsymbol{D}_P^{\boldsymbol{u}})^{**}(\nabla p'')_P, \tag{3.125}$$

which can be shown similar to Equation (3.106). Using this with the Rhie-Chow interpolation in the same way as in the derivation of Equation (3.113) gives a new pressure correction equation

$$\sum_{f \in F(P)} (-\rho_f \overline{\boldsymbol{D}_f^u} \nabla p_f'' \cdot \boldsymbol{S}_f) =$$

$$- \sum_{f \in F(P)} \dot{m}_f^{***} + \sum_{f \in F(P)} (\rho_f \overline{\boldsymbol{H}_f [\boldsymbol{u}'']} \cdot \boldsymbol{S}_f). \qquad (3.126)$$

The underlined parts of Equations (3.125) and (3.126) are again neglected. At the end of the second step all the fields are again updated similar to Equations (3.120)

$$\boldsymbol{u}_P^{****} = \boldsymbol{u}_P^{***} + \boldsymbol{u}_P'' \qquad \text{with } \boldsymbol{u}_P'' = -(\boldsymbol{D}_P^u)^{**}(\nabla p'')_P \qquad (3.127)$$

$$\dot{m}_f^{****} = \dot{m}_f^{***} + \dot{m}_f'' \qquad \text{with } \dot{m}_f'' = -\rho_f \boldsymbol{D}_f^u \nabla p_f'' \cdot \boldsymbol{S}_f \qquad (3.128)$$

$$p_P^{**} = p_P^* + p_P'' \qquad (3.129)$$

This second corrector step can be repeated multiple times, recovering successively additional parts of the originally neglected term $\boldsymbol{H}_P[\boldsymbol{u}']$. One step in the **PISO** algorithm can be summarized as follows

1. Start with a given solution $(\boldsymbol{u}^n, \dot{m}^n, p^n)$ at iteration n as initial guess.

2. Do one step of the **SIMPLE** algorithm to get $(\boldsymbol{u}^{**}, \dot{m}^{**}, p^*)$

3. Assemble a new momentum equation using these updated fields and solve it explicitly to get $\boldsymbol{u}^{***}$

4. Use this field to get a new mass flux field $\dot{m}^{***}$

5. Construct a new pressure correction equation (Eq. (3.126)) with the updated mass flow rate and solve it to get a pressure correction field p''.

6. Update the velocity and pressure fields using this pressure correction with Equation (3.120) to get the corrected fields $(\boldsymbol{u}^{****}, \dot{m}^{****}, p^{**})$.

7. Repeat from step 3 for a predefined number of corrector loops taking this updated values for the construction of the next momentum equation

8. Take the momentary solution as a new initial guess and repeat from step 2 until convergence is achieved.

9. Use the converged solution as the exact solution at iteration $n + 1$ (or time $t + \Delta t$) and proceed to the next iteration.

10. Repeat from step 1 until the last iteration or time step is reached.

Contrary to the original **SIMPLE**, the **PISO** algorithm does not necessarily need under-relaxation for stability, because of this extra correction steps [34, §15.7]. When using the Rhie-Chow interpolation one has to pay special attention in transient cases, when under relaxation for the velocity field is used and in treating the body force terms. In all of this cases the interpolation has to be slightly modified to an extended Rhie-Chow interpolation. Details on this topic can be found in [34, §15.9]. In OpenFOAM® both the **SIMPLE** and the **PISO** algorithm are implemented in a unified framework called **PIMPLE**. This framework allows the user to use different versions of these algorithms as described in [34] by using specific keywords ([24]).

3.3.4 Boundary Conditions

The choice and implementation of the boundary conditions is very critical in CFD to ensure a stable and accurate solution procedure. In the case of sloshing only the wall boundary conditions are of interest. The treatment of the boundaries described here is following the derivations in [34, §15.6]. Recalling Section 2.2 one has a no-slip or slip boundary condition for the velocity and no physical boundary condition for the pressure. Both will be addressed here. The starting point, however, is the necessary modification to the Rhie-Chow interpolation at boundary faces. Since no linear interpolation can be done at such a boundary face b the averaging from Equation (3.100) becomes a low-order extrapolation by setting the boundary value equal to the cell value

$$\overline{\phi_b} = \phi_P.$$

That changes the different average values according to

$$\overline{\boldsymbol{u}_b^*} = \boldsymbol{u}_P^*, \tag{3.130}$$

$$\overline{\boldsymbol{D}_b^{\boldsymbol{u}}} = \boldsymbol{D}_P^{\boldsymbol{u}}, \tag{3.131}$$

$$\overline{\nabla p_b^n} = \nabla p_P^n \tag{3.132}$$

and results in a Rhie-Chow interpolation of the form

$$\boldsymbol{u}_b^* = \overline{\boldsymbol{u}_b^*} - \overline{\boldsymbol{D}_b^{\boldsymbol{u}}} \left(\nabla p_b^n - \overline{\nabla p_b^n} \right) \tag{3.133}$$

$$= \boldsymbol{u}_P^* - \boldsymbol{D}_P^{\boldsymbol{u}} \left(\nabla p_b^n - \nabla p_P^n \right). \tag{3.134}$$

For the details on the wall boundary conditions, the momentum equation (Eq. (2.14)) is rewritten in a semi discretized form with the viscous terms expressed as the incompressible stress tensor $\boldsymbol{\tau}$ and the volume integrals of all the divergence terms transformed to surface integrals

$$\left(\frac{\partial \rho \boldsymbol{u}}{\partial t} \right)_P |V_P| + \sum_{f \in F(P)} (\dot{m}_f \boldsymbol{u}_f) =$$

$$- \sum_{f \in F(P)} (p_f \boldsymbol{S}_f) + \sum_{f \in F(P)} (\boldsymbol{\tau}_f \cdot \boldsymbol{S}_f) + \boldsymbol{f}_P |V_P| . \tag{3.135}$$

All terms containing face values have to be considered at a boundary face. Therefore the sums are split into their interior ($F_{int}(P) := \{\text{inner faces of cell } P\}$) and boundary faces. Without loss of generality we can assume that there is only one boundary face (b) at the cell under consideration. This gives

$$\sum_{f \in F(P)} (\dot{m}_f \boldsymbol{u}_f) = \sum_{f \in F_{int}(P)} (\dot{m}_f \boldsymbol{u}_f) + \dot{m}_b \boldsymbol{u}_b,$$

$$\sum_{f \in F(P)} (p_f \boldsymbol{S}_f) = \sum_{f \in F_{int}(P)} (p_f \boldsymbol{S}_f) + p_b \boldsymbol{S}_b,$$

and

$$\sum_{f \in F(P)} (\boldsymbol{\tau}_f \cdot \boldsymbol{S}_f) = \sum_{f \in F_{int}(P)} (\boldsymbol{\tau}_f \cdot \boldsymbol{S}_f) + \boldsymbol{\tau}_b \cdot \boldsymbol{S}_b$$

$$= \sum_{f \in F_{int}(P)} (\boldsymbol{\tau}_f \cdot \boldsymbol{S}_f) + \boldsymbol{F}_b.$$

All of the boundary values need to be defined. That means they have to be either directly determined by the physical boundary condition or constructed in a way that they are compatible with this condition. The no-slip boundary condition implies that the velocity $\boldsymbol{u}$ is equal to the wall velocity $\boldsymbol{u}_{wall}$. Although this seems like a Dirichlet boundary condition, it is not really one. In fact this means that one has to ensure a zero normal boundary flux while accounting for the right shear stresses. This is done by making the shear stress tangential to the wall and defining the corresponding velocity in a compatible way. $\boldsymbol{F}_b$, which is the force coming from the wall and acting on the fluid, can be divided into a normal and a tangential part

$$\boldsymbol{F}_b = \boldsymbol{F}_\| + \boldsymbol{F}_\perp.$$

The normal force should vanish and therefore

$$\boldsymbol{F}_b = \boldsymbol{F}_\| = \tau_{wall} |S_b|$$

with the shear stress

$$\tau_{wall} = -\mu \frac{\partial \boldsymbol{u}_\|}{\partial d_\perp} \tag{3.136}$$

exerted by the wall [34, §16.6.1.1]. The term $\boldsymbol{u}_\|$ in this equation is the velocity parallel to the wall and $d_\perp$ is the normal distance between the centroid of the boundary element and the wall. It is evaluated according to

$$d_\perp = \boldsymbol{d}_{Pb} \cdot \boldsymbol{n}_w = \frac{\boldsymbol{d}_{Pb} \cdot \boldsymbol{S}_b}{|S_b|} \tag{3.137}$$

where $\boldsymbol{n}_w$ means the unit normal vector to the wall. One can now express the tangential velocity as a difference between the velocity $\boldsymbol{u}$ and its normal component

$$\boldsymbol{u}_\| = \boldsymbol{u} - (\boldsymbol{u} \cdot \boldsymbol{n}_w)\boldsymbol{n}_w = \begin{pmatrix} u_1 - (u_1 n_{w1} + u_2 n_{w2} + u_3 n_{w3})n_{w1} \\ u_2 - (u_1 n_{w1} + u_2 n_{w2} + u_3 n_{w3})n_{w2} \\ u_3 - (u_1 n_{w1} + u_2 n_{w2} + u_3 n_{w3})n_{w3} \end{pmatrix}.$$
$$\text{(3.138)}$$

By approximating the gradient in Equation (3.136) with a linear profile and using Equation (3.138) we can write the wall shear stress as

$$\tau_{wall} \approx -\mu_b \frac{(\boldsymbol{u}_P - \boldsymbol{u}_b)_\|}{d_\perp} = -\mu_b \frac{(\boldsymbol{u}_P - \boldsymbol{u}_b) - ((\boldsymbol{u}_P - \boldsymbol{u}_b) \cdot \boldsymbol{n}_w)\boldsymbol{n}_w}{d_\perp}$$
$$\text{(3.139)}$$

which gives the boundary force

$$\boldsymbol{F}_b \approx -\mu_b |S_b| \frac{(\boldsymbol{u}_P - \boldsymbol{u}_b) - ((\boldsymbol{u}_P - \boldsymbol{u}_b) \cdot \boldsymbol{n}_w)\boldsymbol{n}_w}{d_\perp}. \qquad \text{(3.140)}$$

In the case of a slip boundary condition there is no wall shear stress which results in a vanishing boundary force $\boldsymbol{F}_b$. In our case a wall is defined in a way so that the fluids can not go through it. Therefore the mass flux at any wall $\dot{m}_b$ is also zero. Since there is no condition on the pressure at the boundary face there can not be a defined value. Additionally we have no mass flux at a wall which means that there is no need for a flux correction in the pressure equation ($\dot{m}_b' = 0$, cf. Equation (3.103)). However, a pressure value has to be defined at the boundary. This is done by extrapolation. Relatively simple ways for that are either a low-order extrapolation

$$p_b = p_P \qquad \text{(3.141)}$$

or a truncated Taylor series expansion

$$p_b = p_P + \nabla p_P \cdot \boldsymbol{d}_{Pb}. \qquad \text{(3.142)}$$

For a closed vessel like the geometries under consideration every boundary face is of the type wall. This also means that there is no defined pressure value at any boundary. The pressure is also only present in the incompressible momentum equations through its gradient. In this case no meaningful absolute pressure level can be calculated, only pressure differences. Furthermore the coefficient matrix of the resulting linear system after discretization will be singular. To get a regular matrix one can define a pressure value at one point in the computational domain. The pressure field is then calculated relative to this point ([34, §15.6.2.5]).

3.4 The VOF Method

Following the description in Section 2.3, the distribution of the free surface is described by the phase function H and the corresponding transport equation (Eq. (2.50)). The discretization of this equation is done with the volume of fluid method (VOF, originally introduced in [23]). It will be described here in general according to [55] and following the respective implementation in OpenFOAM® as given in [44], [2] and [12]. For that the discontinuous phase function H is replaced by a phase fraction field α. This α field takes values between 0 and 1 and describes the fraction of the reference fluid at a point in space. In the discrete case this means for any cell the fraction of the cell volume occupied by the reference fluid. It is then used to define the fluid properties analogously to the phase function in Equation (2.51). In the original VOF approach the advection of the volume fraction field follows a three step procedure. Starting from an old α field and a given velocity field the first step is the reconstruction of the interface in every cell with values $0 < \alpha_P < 1$. Several methods for that have been proposed in literature apart from the one adopted in [23]. Basic ones are the **SLIC** (simple line interface calculation,[35]) or **PLIC** (piecewise linear interface calculation,[60]) method. Some of these algorithms require the computation of the interface normal. Examples for the different methods used for that purpose are the Youngs Finite Difference Method ([61]), the **ELVIRA** method (efficient least-squares VOF interface reconstruction algorithm, [42]) and the least-squares fit method ([46]). This is followed by the advection of the reconstructed surface. A good overview of different options for this step is given in [55, §5.4]. The last step is the calculation of the advected α field based on the moved surface. This new α field is then used in the Navier-Stokes equations for the computation of the mixture properties and the flow field. The implementation used in OpenFOAM® follows a different approach. There is no geometric reconstruction and separate advection. The evolution of the α field is computed by using the general conservation equation

$$\frac{\partial \alpha}{\partial t} + \nabla \cdot (\boldsymbol{u}\alpha) = 0. \tag{3.143}$$

This equation can be discretized like any scalar transport equation (c.f. §3.2). However, one has to ensure that there is no extensive smoothing of the gradient at the free surface by the used discretization methods. Optimally one wishes to keep the gradient zone (the cells with values between 0 and 1) restricted to a width of just a few cells while it is necessary to assure that the values are bounded between zero and one. Therefore an extended version of this equation is used:

$$\frac{\partial \alpha}{\partial t} + \nabla \cdot (\boldsymbol{u}\alpha) + \nabla(\boldsymbol{u}_r\alpha(1-\alpha)) = 0. \tag{3.144}$$

It can be derived by defining separate transport equations for α and $(1-\alpha)$

$$\frac{\partial \alpha}{\partial t} + \nabla \cdot (\boldsymbol{u}_\alpha\alpha) = 0, \tag{3.145}$$

$$\frac{\partial (1-\alpha)}{\partial t} + \nabla \cdot (\boldsymbol{u}_{1-\alpha}(1-\alpha)) = 0 \tag{3.146}$$

with their respective velocity fields $\boldsymbol{u}_\alpha$ and $\boldsymbol{u}_{1-\alpha}$. Defining the velocity of the mixture as a weighted average

$$\boldsymbol{u} = \alpha\boldsymbol{u}_\alpha + (1-\alpha)\boldsymbol{u}_{1-\alpha},$$

Equation (3.145) can be rearranged to Equation (3.144) with the compression velocity defined as $\boldsymbol{u}_r := \boldsymbol{u}_\alpha - \boldsymbol{u}_{1-\alpha}$. The additional term is an artificial compression term ([44], [12]). This term is only active after discretization and in the vicinity of the interface due to the product $\alpha(1 - \alpha)$ which is zero for any empty or full cell. It does not affect the original continuous equation, because the indicator function at any point is either 0 or 1 but nothing in between. The compression velocity $\boldsymbol{u}_r$ has to be defined in a suitable way to ensure a sharp interface region ([44]). It is computed in the discrete system via face fluxes. The transport equation is solved explicitly and the desired boundedness is achieved by limiting the calculated face fluxes with the so-called **MULES** limiter (Multidimensional Universal Limiter for Explicit Solution, [12]) which is a special case of the **FCT** (Flux-Corrected Transport, [3], [62], [27]) methodology. In this method the face fluxes are limited (in contrary to the face values in case of **HR** schemes). The first step is to integrate Equation (3.144) over a cell P and using Gauss Theorem and a one point numerical integration similar to the discretization of the convection terms in section 3.2.2. This gives a semi-discretized equation of the form

$$\left(\frac{\partial \alpha}{\partial t}\right)_P |V_P| + \sum_f \alpha_f \boldsymbol{u}_f \cdot \boldsymbol{S}_f + \sum_f \alpha_f (1 - \alpha)_f (\boldsymbol{u}_r)_f \cdot \boldsymbol{S}_f = 0. \quad (3.147)$$

Using the explicit Euler temporal scheme and defining the face fluxes $F_f := \boldsymbol{u}_f \cdot \boldsymbol{S}_f$ and $F_f^r := (\boldsymbol{u}_r)_f \cdot \boldsymbol{S}_f$ this becomes

$$\frac{\alpha_P^{n+1} - \alpha_P^n}{\Delta t} |V_P| + \sum_f \alpha_f^n (F_f)^n + \sum_f \alpha_f^n (1 - \alpha)_f^n (F_f^r)^n = 0. \quad (3.148)$$

The face values α_f are calculated with any high resolution scheme (e.g. the **TVD** conforming Van Leer scheme) as in section 3.2.2 whereas the volumetric flux F_f is taken from the pressure-velocity coupling in the **PISO** algorithm ($\dot{m}_f = \rho \boldsymbol{u}_f \cdot \boldsymbol{S}_f = \rho F_f$). The compressive flux is computed as

$$F_f^r := \hat{n}_f C_\alpha \left| \frac{F_f}{S_f} \right| \quad (3.149)$$

with a compression constant C_α and the face unit normal flux $\hat{n}_f$. This flux is defined using the gradient of the phase fraction

$$\hat{n}_f = \frac{(\nabla\alpha)_f}{|(\nabla\alpha)_f + \delta n|} \cdot \boldsymbol{S}_f \quad (3.150)$$

where δn is a small stabilization factor to avoid division by zero which is calculated according to

$$\delta n = \frac{\varepsilon}{\left(\frac{1}{N}\sum_{i=1}^{N}|V_i|\right)^{\frac{1}{3}}}. \quad (3.151)$$

In OpenFOAM® ε is set to 10^{-8}. This computations are followed by the application of a **FCT** technique. This is done by combining the fluxes and the face values to a total high order face flux

$$F_f^{\alpha,H} = \alpha_f F_f + \alpha_f(1-\alpha)_f F_f^r.$$

Equation (3.148) can then be rearranged with these combined face fluxes to get

$$\alpha_P^{n+1} = \alpha_P^n - \frac{\Delta t}{|V_P|}\sum_f (F_f^{\alpha,H})^n. \quad (3.152)$$

Additionally a low order flux

$$F_f^{\alpha,L} = \alpha_f^L F_f$$

is computed with the Upwind Differencing Scheme. While directly using the upwind flux $F_f^{\alpha,L}$ in Equation (3.152) would keep the solution bounded it is too diffusive and would smear out the gradient. $F_f^{\alpha,H}$ on the other hand keeps the gradient sharp at the expense of possible unboundedness. For **FCT** the low order fluxes are then subtracted from the ones with higher order to get the anti-diffusive flux

$$A_f = F_f^{\alpha,H} - F_f^{\alpha,L}, \quad (3.153)$$

which is then limited to get the corrected flux

$$F_f^{\alpha,C} = F_f^{\alpha,L} + \lambda_f A_f. \quad (3.154)$$

The λ_f are suitable weighting factors between 0 and 1 to ensure that no new local extrema are created and the global limits are fulfilled. The resulting limited fluxes are then inserted in Equation (3.152) to get

$$\alpha_P^{n+1} = \alpha_P^n - \frac{\Delta t}{|V_P|} \sum_f (F_f^{\alpha,C})^n \tag{3.155}$$

which is an explicit equation for the α field at the new time step. The crucial step in this procedure is the calculation of the weighting factors λ_f. In the **MULES** limiter this is done by an iterative manner ([12]). According to [27], this iteration converges to an analytical value given in [62]. It starts with the calculation of the local extrema

$$\alpha_P^{maxL,n} = \max\{\alpha_P^n, \alpha_N^n\} \qquad N \in NB(P),$$
$$\alpha_P^{minL,n} = \min\{\alpha_P^n, \alpha_N^n\} \qquad N \in NB(P)$$

and the total inflow F_P^+ and outflow F_P^- of the antidiffusive flux A

$$F_P^+ = -\sum_f A_f^-,$$
$$F_P^- = \sum_f A_f^+$$

where A_f^- are the inflows and A_f^+ are the outflows per face. This is followed by a correction of the local maxima/minima with respect to the global limits α^{maxG} and α^{minG}

$$\alpha_P^{max,n} = \min\{\alpha^{maxG}, \alpha_P^{maxL,n}\},$$
$$\alpha_P^{min,n} = \max\{\alpha^{minG}, \alpha_P^{minL,n}\}.$$

These values are then used in combination with the low-order flux to compute the net flux per cell

$$Q_P^+ = \frac{V}{\Delta t}(\alpha_P^{max,n} - \alpha_P^n) + \sum_f F_f^{\alpha,L},$$
$$Q_P^- = \frac{V}{\Delta t}(\alpha_P^n - \alpha_P^{min,n}) - \sum_f F_f^{\alpha,L}.$$

The iteration starts with the initial guess $\lambda_f^1 = 1.0$ for every face. With λ_f^ν as the value from a previous iteration, first two intermediate values are computed for every cell

$$
\lambda_P^{+,\nu+1} = \max\left\{\min\left(\frac{-\sum_f \lambda_f^\nu A_f^- + Q_P^-}{F_P^-}, 1\right), 0\right\},
$$

$$
\lambda_P^{-,\nu+1} = \max\left\{\min\left(\frac{+\sum_f \lambda_f^\nu A_f^+ + Q_P^+}{F_P^+}, 1\right), 0\right\}.
$$

These are then used to compute the new face values

$$
\lambda_f^{\nu+1} = \begin{cases} \min\{\lambda_P^{+,\nu+1}, \lambda_N^{-,\nu+1}\}, & \text{if } A_f \geq 0, \\ \min\{\lambda_P^{-,\nu+1}, \lambda_N^{+,\nu+1}\}, & \text{if } A_f < 0. \end{cases} \tag{3.156}
$$

Here N is the neighbor cell corresponding to face f. This loop is repeated for a predefined number of iterations. The whole solution procedure for the phase fraction equation is explicit (Eq. (3.155)), which gives an additional limitation on the maximal usable time step (cf. section 3.2.4). For that reason temporal subcycling is often used in the solution of Equation (3.144). That means the total time step is divided into a number of smaller intervals, which are then solved successively in the algorithm for the α equation without recomputing the complete flow field after every sub-step ([2] and [12]).

Due to the one-fluid formulation only one pressure field is present in the model. However, the gradient normal to a stationary wall at a contact point of the free surface is different for each phase due to the different densities of the fluids [2]. For a simpler definition of boundary conditions and a more stable solution a modified pressure is introduced

$$
p_{rgh} = p - \rho \boldsymbol{g} \cdot \boldsymbol{x} \tag{3.157}
$$

with the gravitational acceleration $\boldsymbol{g}$. The negative pressure gradient is then expressed as in

$$
-\nabla p = -\nabla p_{rgh} - \boldsymbol{g} \cdot \boldsymbol{x} \nabla \rho - \rho \boldsymbol{g}. \tag{3.158}
$$

Rearranging Equation (3.158) results in

$$-\nabla p + \rho \boldsymbol{g} = -\nabla p_{rgh} - \boldsymbol{g} \cdot \boldsymbol{x} \nabla \rho. \qquad (3.159)$$

Taking the gravitational force out of the body force term in the momentum equation

$$\boldsymbol{f} = \tilde{\boldsymbol{f}} + \rho \boldsymbol{g}$$

the whole left side of Equation (3.159) can be substituted by the right hand side in the momentum equation. This results in the following modified momentum equation

$$\int_{V_P} \frac{\partial \rho \boldsymbol{u}}{\partial t} \, \mathrm{d}\boldsymbol{x} + \int_{V_P} \nabla \cdot \{\rho \boldsymbol{u} \boldsymbol{u}\} \, \mathrm{d}\boldsymbol{x} =$$

$$-\int_{V_P} \nabla p_{rgh} \, \mathrm{d}\boldsymbol{x} - \int_{V_P} \boldsymbol{g} \cdot \boldsymbol{x} \nabla \rho \, \mathrm{d}\boldsymbol{x} +$$

$$\int_{V_P} \nabla \cdot \{\mu \nabla \boldsymbol{u}\} \, \mathrm{d}\boldsymbol{x} + \int_{V_P} \nabla \cdot \{\mu (\nabla \boldsymbol{u})^T\} \, \mathrm{d}\boldsymbol{x} + \int_{V_P} \tilde{\boldsymbol{f}} \, \mathrm{d}\boldsymbol{x} \qquad (3.160)$$

which is then used in the PISO algorithm ([2],[12],[44]).

3.5 Summary of the solution algorithm

With the results of this chapter one step of the solution algorithm forward in time can be summarized in the following way:

1. Start with the solution of the previous time step. This solution consists of the velocity field $\boldsymbol{u}^n$, the modified pressure field p_{rgh}^n, the mixture density ρ^n and the viscosity μ^n and the phase fraction field α^n, which are all defined on the cell centers. Additionally the volume flux F_f^n and the mass flux $\dot{m}_f^n$ are stored on the face centers.

2. Compute the evolution of the phase fraction field as described in Section 3.4 using the **MULES** limiter and temporal sub-cycling. Use the new α field to calculate the updated density and viscosity fields.

3. Take the updated fields and assemble the momentum and pressure equations. Use the solution procedure of the **PISO** algorithm (sections 3.3.2 and 3.3.3) to compute the pressure, velocity and flux fields for the new timestep

4. Take the computed fields as a new starting point and repeat this sequence for the next time step

4 Validation and Application to Sloshing

The numerical solution procedure described in the previous chapters is finally applied to two different problems. Both are sloshing problems. The first is a simple validation test case and used to find a suitable combination of discretization schemes and parameter settings to achieve an accurate solution. The second test case is a real fuel tank exposed to a given acceleration profile. This model was chosen to show the usability of the presented method for the simulation of real world sloshing problems and the used pre- and postprocessing method to get results with respect to sloshing acoustics. As mentioned in the beginning the OpenFOAM® software (version v1712, [36]) is used for these test cases. The solution algorithm from the previous chapter is implemented in the two solvers interFoam and interDyMFoam. The difference between the two of them is that the interDyMFoam implementation can deal with changing (dynamic) meshes. This capability is necessary if one wants to use the rigid body motion approach for the displacement of the tank. The interFoam solver is only able to apply the outer displacement through body forces in a non-inertial frame of reference.

4.1 Sloshing Box

The first test case is a simple rectangular box. It is based on an experimental test done at the Technical University of Madrid ([4], [14], [49] and [48]). The box has the following dimensions:

- Length: $900mm$

© Springer Fachmedien Wiesbaden GmbH, part of Springer Nature 2019
M. Jäger, *Fuel Tank Sloshing Simulation Using the Finite Volume Method*,
BestMasters, https://doi.org/10.1007/978-3-658-25228-1_4

- Width: $62mm$

- Height: $508mm$

It is partially filled with water up to a filling height of $93mm$ at rest.
The material parameters for this test case are defined in Table 4.1.
The sloshing is induced by a periodic rotational displacement with

	water	air
density (ρ)	$998\frac{kg}{m^3}$	$1.2089\frac{kg}{m^3}$
kinematic viscosity (ν)	$8.96 \cdot 10^{-07}\frac{m^2}{s}$	$1.50195 \cdot 10^{-05}\frac{m^2}{s}$
surface tension coefficient (σ)	$0.072\frac{m}{s^2}$	

Table 4.1: Material properties

its axis in the middle of the bottom wall (see Figure 4.1). The rotational displacement is starting at rest and ramps up until it reaches
a periodic situation with an amplitude of approximately 4 degrees
and a period time of approximately 1.63 seconds (see Fig. 4.2). A
pressure sensor was installed at the point P_1 which is located in the
middle of the left wall at a height of $93mm$ (the water level). The
pressure variation at this point is also recorded in the simulation and
then compared to the measured values from the experimental setup.
Additionally the distribution of the free surface is compared to the
experiment at several instances. The simulation time for each run is
set to 8 seconds of physical time.
The simulations are done with the interDyMFoam solver and the
motion of the tank is applied via the rigid body motion approach.
The base mesh for this model is a pure hexahedral mesh with 238680
elements (180 over the length, 13 over the width and 102 for the
height). This results in a cell size (edge length) of approximately
$5mm$ (see Fig. 4.3). The temporal discretization is done with a
uniform fixed time step of $0.0001s$. As time scheme the implicit Euler
scheme is used and the convection terms are discretized with the Van
Leer scheme. The constant C_α in the phase fraction equation is set to

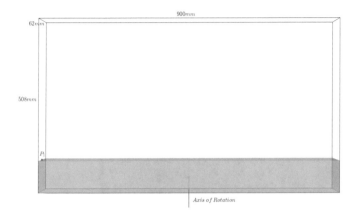

Figure 4.1: Geometry of the box

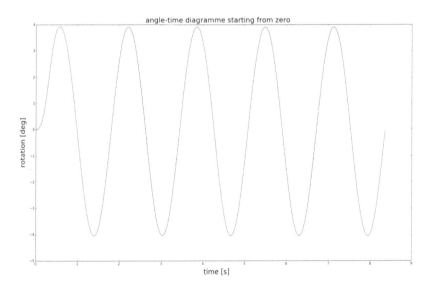

Figure 4.2: Rotational angle vs time

1.0 and the iteration for the λ values in the MULES limiter is done three times. The systems of linear equations are solved with a GAMG solver.

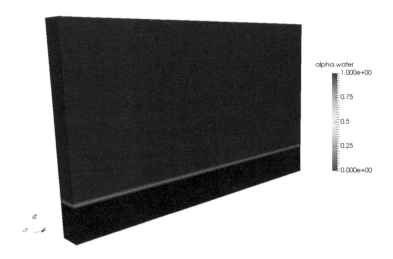

Figure 4.3: Base mesh with filling level

The obtained pressure values are in good agreement with the experimental data, particularly for the first impact of a wave (see Fig. 4.4). However, there is an increasing discrepancy between them with ongoing time. In many cases, especially for sloshing acoustics, it is only necessary to evaluate the fluid flow for about 5 seconds. Therefore this model with the respective settings is found to be a good starting point. The following tests are mainly compared using only the pressure values for a small time interval between 2 and 3 seconds of physical time (the first peak) since this is the most important time span.

Based on this model some numerical tests are carried out in order to identify the influence of different parameters on the simulated sloshing behavior. The first tests consist of a comparison of different time step settings where both fixed and adjustable time intervals are used

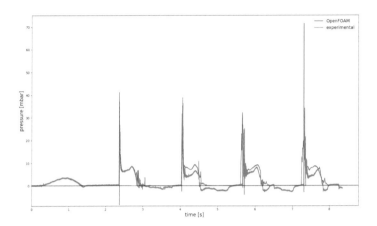

Figure 4.4: Comparison of the experimental pressure with the basic test case

(see Tab. 4.2). The adjustable time step is based on a defined maximum Courant (CFL) number. Although the implicit Euler scheme is unconditionally stable the maximum admissible Courant number is theoretically bounded by 1 since the time integration in the **MULES** limiter is explicitly done with the forward Euler scheme. Varying the

Test	(initial) Δt	adj./fixed.	max CFL Nr.	max Δt
Time Step 1	0.0002	fixed	-	-
Time Step 2	0.0001	fixed	-	-
Time Step 3	0.00005	fixed	-	-
Time Step 4	0.0001	adjustable	0.5	0.005
Time Step 5	0.0001	adjustable	1.0	0.005
Time Step 6	0.0001	adjustable	1.0	0.01

Table 4.2: Settings for the different time step tests

fixed time step sizes does no big change in the prediction of the peaks.

Only the peak height is getting a little bit higher with smaller step sizes (Fig. 4.5). A comparison of the base case with the 3 settings for the adjustable time step size shows that *Time Step 4* and *Time Step 5* are in better agreement with the data than the others. While *Time Step 5* gives a better approximation of the peak height, *Time Step 4* is reproducing the temporal occurrence of the wave impact more accurately (see Fig. 4.6). Due to this and the fact that a lower Courant number reduces the risk of a diverging simulation the combination of test *Time Step 4* is found to be more suitable for this case.

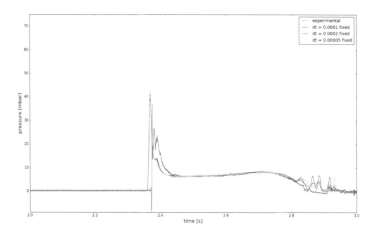

Figure 4.5: Comparison of different fixed time step sizes

The time step comparison is followed by an examination of different time discretization schemes (3.2.4). Both the implicit Euler and the Crank-Nicolson schemes are tested. However, the Crank-Nicolson scheme is only used blended with the Euler scheme, because the usage of a pure Crank-Nicolson time integration is not recommended due to possible stability problems (no boundedness, see [25]). The tested settings are depicted in Table 4.3. For the given time step size and spatial discretization there is no significant difference recognizable between the 3 Tests. Only the Crank-Nicolson scheme with a blending

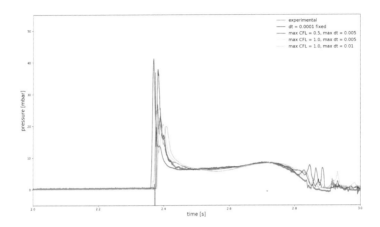

Figure 4.6: Comparison of different settings for the adjustable time step

	ddtScheme	blending coefficient
Time Scheme 1	Euler	0
Time Scheme 2	Crank-Nicolson	0.5
Time Scheme 3	Crank-Nicolson	0.9

Table 4.3: Different temporal discretization schemes

factor of 0.9 exhibits a more oscillatory behavior, especially in the smooth region after the pressure peak (see Fig. 4.7).

For all the further comparisons from here on the temporal discretization is changed to an adjustable time step with a maximal CFL number of 0.5 and a maximum time step size of 0.005 (see test Time Step 4, Tab. 4.2). Additionally the temporal discretization scheme is changed to a blended Crank-Nicolson and Euler scheme with a blending factor of 0.5 (see Tab. 4.3, Time Scheme 2). This scheme is chosen over the pure Euler because of its theoretically higher accuracy. The next test is a comparison of different mesh sizes for the influence

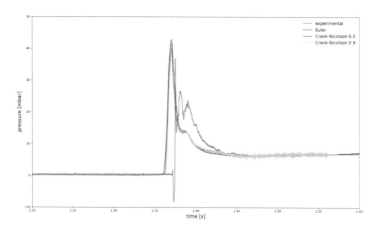

Figure 4.7: Comparison of different time discretization schemes

of the spatial resolution. The used meshes are summarized in Table 4.4. From this results it can be seen that the decreasing mesh density

test case	cell size $[mm]$	number of elements
Mesh 1	~ 10	27540
Mesh 2	~ 5	238680
Mesh 3	~ 2.5	1.836e6

Table 4.4: Properties of the different meshes

is shifting the first peak forward in time (Fig. 4.8). An interesting fact is that the finest mesh does exhibit a very similar curve to the experimental data at the peak, but with a higher maximum pressure value. However, from the given data it is not really possible to draw any conclusion from this since the experimental values are exhibiting severe oscillations. Nevertheless, the setting with 5 mm cell size is giving acceptable results to be used for bigger models where a smaller cell size would result in too high cell counts.

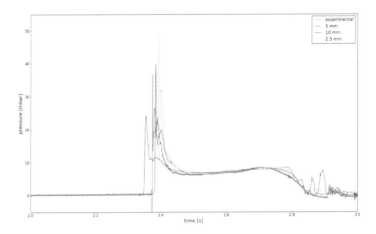

Figure 4.8: Comparison of different cell sizes

The fourth test is a comparison of the different discretization schemes for the convection terms described in Section 3.2.2. The three tests are done with the schemes presented in Table 4.5. Interpolating the face

test case	convection scheme
Conv 1	**UD**
Conv 2	**SOU**
Conv 3	Van Leer

Table 4.5: Different Convection Schemes

values with the linear interpolation of **CD** did not give the desired results because was too dispersive and the simulation diverged after 1 second of simulated time. Therefore only the other three schemes are used in the comparison. From that we see that the Van Leer scheme is performing much better than **UD** and **SOU**. Upwind differencing is clearly too diffusive and smoothing out the peaks. Second Order Upwinding is not predicting the time of the wave impact accurately enough and gives a too big and too sharp pressure peak.

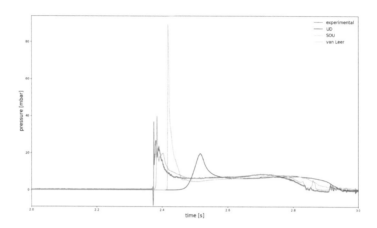

Figure 4.9: Comparison of different convection schemes

For the last test the dependence of the results on the constant C_α from the compression term in the phase fraction equation (Eq. 3.144) is examined. The different values used for that are given in Table 4.6. These results show that a smaller value for C_α can lead to a

test case	C_α
C_α 1	0.25
C_α 2	0.5
C_α 3	1.0
C_α 4	2.0

Table 4.6: Different C_α values

reduction of the peak value. Between the values 1 and 2 there is no major difference. Therefore a value of 1 is assumed to be sufficient. The results of the previous tests recommend a combination of the

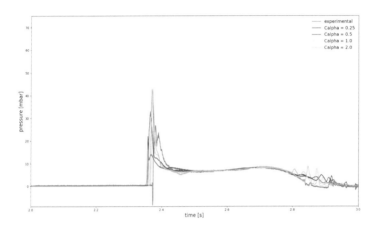

Figure 4.10: Comparison of different C_α values

blended Crank-Nicolson scheme, adjustable time stepping with a maximum CFL number not higher than 0.5, the Van Leer limiter for the convection terms and a value of 1 for C_α. Using these settings results in good conformance for the first peak, although there is a short shift in time left (Fig. 4.11).
For this improved configuration a visual comparison to the real test for some time instances at the second impact on the left wall is done. The figures show situations shortly before the impact (Fig. 4.12), the moment of the impact (Fig. 4.13) and after the impact (Fig. 4.14). Comparing the evolution of the free surface for this test case in combination with the relatively good approximation of the pressure curve shows that it is a good approximation of the real flow. These settings are therefore chosen to be accurate enough for the usage on real tanks.

4.2 Fuel Tank

This second numerical test is done on a real automotive fuel tank geometry. Three variants of the tank are used. One consists only

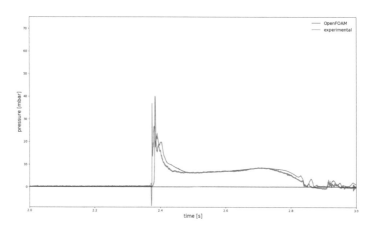

Figure 4.11: Pressure comparison with changed settings

Figure 4.12: Comparison of the free surface before the impact

Figure 4.13: Comparison of the free surface at the impact

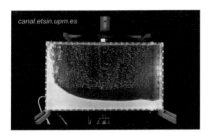 

Figure 4.14: Comparison of the free surface after the impact

of the tank hull while the second model includes also all the inner components of a tank system. In the third variant there are additional anti-sloshing devices (baffles) installed. All of them are asymmetrically filled with 45 liters of water (see Fig. 4.15). This asymmetric filling is chosen because it resembles the situation after a fresh refueling of the tank better where only the main chamber gets filled first until the fuel reaches the saddle point in the middle. The material properties

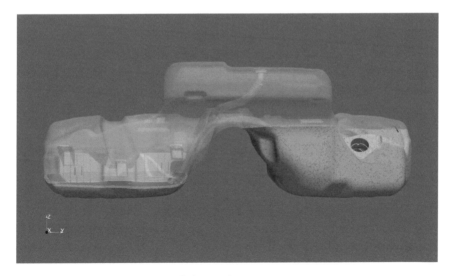

Figure 4.15: Initial filling of the tank

in this case are the same as in the previous case (Table 4.1). The sloshing is induced by a start-stop driving profile (see Fig. 4.16) The

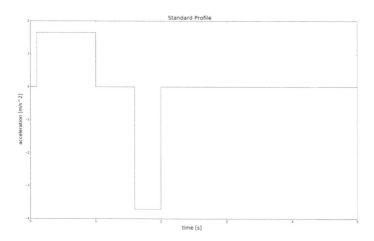

Figure 4.16: Acceleration vs time for the start-stop profile

meshing is done with a cell size of $5mm$ in the main fluid volume and
$\sim 1mm$ at the boundaries and the inner parts (see Figure 4.17 for
the geometry of the full model with all parts and Figure 4.18 for the
mesh). This gives the following cell numbers:

- only tank hull: 1595243 cells

- inner parts without baffles: 2470680 cells

- inner parts with baffles: 2751513 cells

Due to the huge amount of data it is not feasible to write all the results
with a fine enough writing interval. But the accurate recognition of
pressure peaks needs a small writing interval for the output. Additionally it would be very time consuming to directly evaluate the pressure
distribution over the boundaries. Therefore in the pre-processing step
the whole tank hull is split into several smaller sectors. This can
be either manually defined areas of interest or defined by a regular

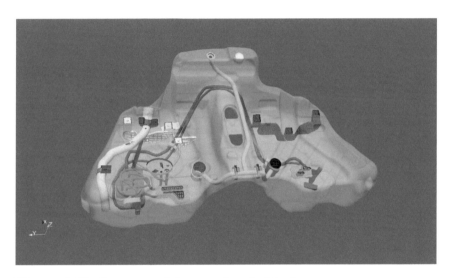

Figure 4.17: Geometry of the complete fuel tank

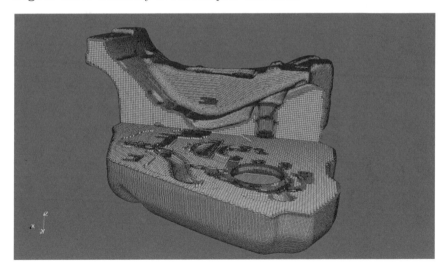

Figure 4.18: Mesh of the complete fuel tank

grid in all coordinate directions. For all this sectors the maximal and average pressure values are written by the solver at every time step or at small enough time intervals for being able to capture wave impacts. These pressure-time curves are then plotted for each sector and used to do a comparison between the different geometries or separately examine them to find critical peaks. With the desired size/number of such sectors one can control the ratio between accuracy in finding the local wave impacts and the necessary amount of evaluation in the post-processing. A splitting of the shell in $100mm \times 100mm$ parts is normally enough to identify the critical waves. In this test case the splitting was done manually and with a coarser non-regular distribution of the sectors. This resulted in 30 pressure monitoring surfaces (Fig. 4.19).

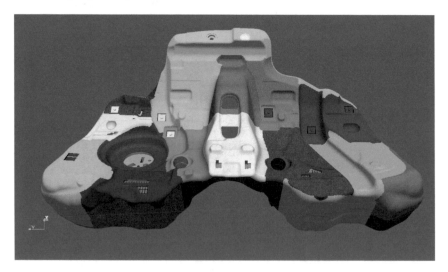

Figure 4.19: Splitting of the tank hull

An analysis of the plots shows different phenomena. In some areas the two models with inner parts exhibit reduced pressure peaks compared to the hull alone while in other regions the complete model leads to new impacts. In the following examples the data related to the pure hull model is colored in blue while the data from the model without baffles is red and that from the full model is green. The first sector examined is sector 1, which is the most right brown sector in Figure 4.19. This sector shows the expected reduction in peak height. Only some short peaks resulting from the impact of small water drops are apparent (see Fig. 4.20). Because of their small scale this short peaks have to be ignored. This is one of the problems by plotting the maximal pressure over the surface because it does not filter such occurrences. This reduction can also be seen in the visualization of the free surface for the same area at a simulated time of $3.35s$. A reduction can also be found for sector 12 (the upper part of the higher turquoise area in Fig. 4.19). In this case a big difference between the blue and the green/red curves is obvious (see Fig. 4.22). This is because the liquid does not reach these upper parts with the same intensity after adding the inner parts. A contrary situation can be seen for sector 18 (the small red sector on the left in Fig. 4.19), where the model with the additional baffles results in a new strong pressure peak (see Fig. 4.23).

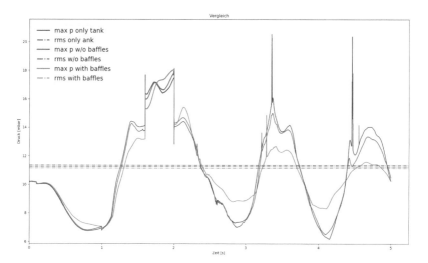

Figure 4.20: max pressure on sector 1

These are only three examples of the 30 pressure plots for this test

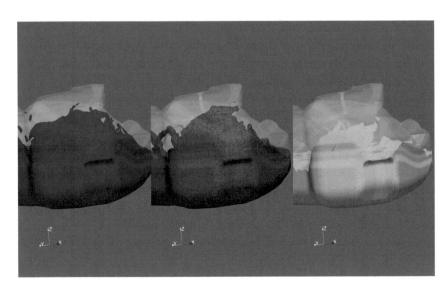

Figure 4.21: comparison of the free surface for sector 1

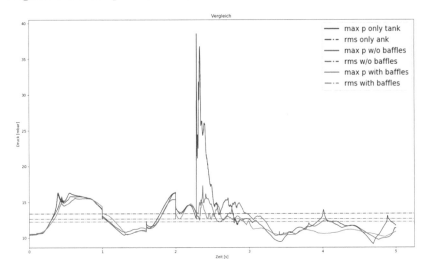

Figure 4.22: max pressure on sector 12

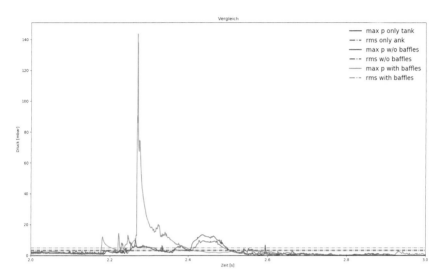

Figure 4.23: max pressure on sector 18

case. For a really useful investigation the sectors would have to be smaller, which can result in a significant higher number of plots. The current post-processing procedure for a full examination consists of visually inspecting all the plots of the monitored pressure values, finding critical peaks and investigating them further with a visualization of the flow field to get a better insight into the motion of the respective wave. This methodology has been found to be suitable to identify critical wave motion and therefore possible noise sources due to the impact. The computed pressure field at specific time instances can also be used as surface loads in a structural simulation for any of the inside parts. These structural simulations are sometimes necessary for being able to assure that the parts will not break due to the maximum sloshing loads.

5 Conclusion and Prospect

The model presented in Chapters 2 and 3 is able to describe the
evolution of a free surface between two incompressible fluids. But by
using it one has to keep in mind that there are severe assumptions
necessary to get the respective partial differential equations (§2.6).
The described discretization with the finite volume method and a
volume-of-fluid approach for the interface description is generally
found to be suitable for the simulation of fuel tank sloshing. However,
not every combination of discretization practice leads to satisfactory
results, particularly if the mesh size can not be arbitrary small due to
the overall fluid volume of the geometry. Therefore a good mixture
between accuracy and stability as the one derived for the validation
case at the end of Section 4.1 had to be found. The final setting for this
case is able to accurately predict the occurrence of a wave impact, but
especially the temporal accuracy is decreasing with a longer simulated
time span. Therefore it is more reliable for the first few seconds of
simulated physical time. For the evaluation oft the sloshing in a real
fuel tank an additional post-processing procedure had to be developed.
The developed simulation method with OpenFOAM®, as described in
Section 4.2 is already used in the development process for fuel tanks
at the moment. However, it is permanently under improvement and
there are many possibilities for that. For example, OpenFOAM®
does not only provide the numerical options described in this work.
There are additional methods which can be used to increase accuracy
and reduce simulation time. Particularly the second factor will be an
important point for future research. Methods like the time step sub-
cycling mentioned in Section 3.4 or a semi-implicit **MULES** limiter
are promising to be of great use. Also the influence of turbulence
on the sloshing loads has not been addressed by now. Besides the

© Springer Fachmedien Wiesbaden GmbH, part of Springer Nature 2019
M. Jäger, *Fuel Tank Sloshing Simulation Using the Finite Volume Method*,
BestMasters, https://doi.org/10.1007/978-3-658-25228-1_5

numerics there are many other possibilities to improve the model. The inclusion of fluid structure interaction between the fuel and the structural parts would be beneficial for the prediction of damage due to the sloshing. An important aim for the future is the simulation of a refueling process. Because of the then necessary modeling of the complete ventilation system, an accurate model for the phase change between the liquid fuel and its gas phase will be needed.

Important Definitions and Theorems

Here some important definitions or theorems used throughout this work are depicted. All of the definitions are given for the three dimensional space R^3 with the cartesian coordinates

$$\boldsymbol{x} = \begin{pmatrix} x_1 \\ x_2 \\ x_3 \end{pmatrix}. \tag{1}$$

A (column) vector is written with a boldface letter

$$\boldsymbol{f} = \begin{pmatrix} f_1 \\ f_2 \\ f_3 \end{pmatrix}. \tag{2}$$

The transpose of a column vector is the respective row vector

$$\boldsymbol{f}^T = \begin{pmatrix} f_1 & f_2 & f_3 \end{pmatrix}. \tag{3}$$

The dot- or scalar product between two vectors $\boldsymbol{a}$ and $\boldsymbol{b}$ is denoted by

$$\boldsymbol{a} \cdot \boldsymbol{b} = \sum_i a_i b_i. \tag{4}$$

The partial derivative of a function f with respect to x_i is written as $\frac{\partial f}{\partial x_i}$ and the symbol ∇ means the partial derivatives along all the coordinate axes written as a vector

$$\nabla = \begin{pmatrix} \frac{\partial}{\partial x_1} \\ \frac{\partial}{\partial x_2} \\ \frac{\partial}{\partial x_3} \end{pmatrix}, \tag{5}$$

© Springer Fachmedien Wiesbaden GmbH, part of Springer Nature 2019
M. Jäger, *Fuel Tank Sloshing Simulation Using the Finite Volume Method*,
BestMasters, https://doi.org/10.1007/978-3-658-25228-1

which is used to define the gradient of f:

$$\operatorname{grad} f = \nabla f = \begin{pmatrix} \frac{\partial f}{\partial x_1} \\ \frac{\partial f}{\partial x_2} \\ \frac{\partial f}{\partial x_3} \end{pmatrix}. \tag{6}$$

The divergence operator of a vector quantity $\boldsymbol{f}$ is given as the dot product of ∇ with $\boldsymbol{f}$

$$\operatorname{div} f = \nabla \cdot \boldsymbol{f} = \sum_i \frac{\partial f_i}{\partial x_i} \tag{7}$$

and the Lapace operator Δ of a scalar quantity is defined as

$$\Delta f = \nabla \cdot (\nabla f) = \nabla^2 f = \sum_i \frac{\partial^2 f}{x_i^2} \tag{8}$$

Tensors are also written in boldface letters

$$\boldsymbol{T} = \begin{pmatrix} T_{11} & T_{12} & T_{13} \\ T_{21} & T_{22} & T_{23} \\ T_{31} & T_{32} & T_{33} \end{pmatrix} \tag{9}$$

and the transposed form is given as

$$\boldsymbol{T}^T = \begin{pmatrix} T_{11} & T_{21} & T_{31} \\ T_{12} & T_{22} & T_{32} \\ T_{13} & T_{23} & T_{33} \end{pmatrix}. \tag{10}$$

Tensors can be created by the dyadic product of 2 vectors

$$\boldsymbol{ab} = \begin{pmatrix} a_1 b_1 & a_1 b_2 & a_1 b_3 \\ a_2 b_1 & a_2 b_2 & a_2 b_3 \\ a_3 b_1 & a_3 b_2 & a_3 b_3 \end{pmatrix}. \tag{11}$$

Taking the dyadic product of the ∇ operator and a vector-field $\boldsymbol{f}$ gives the gradient

$$\nabla \boldsymbol{f} = \begin{pmatrix} \frac{\partial f_1}{\partial x_1} & \frac{\partial f_2}{\partial x_1} & \frac{\partial f_3}{\partial x_1} \\ \frac{\partial f_1}{\partial x_2} & \frac{\partial f_2}{\partial x_2} & \frac{\partial f_3}{\partial x_2} \\ \frac{\partial f_1}{\partial x_3} & \frac{\partial f_2}{\partial x_3} & \frac{\partial f_3}{\partial x_3} \end{pmatrix}. \tag{12}$$

The dot product of a tensor $\boldsymbol{T}$ and a vector is defined as

$$\boldsymbol{T} \cdot \boldsymbol{f} = \begin{pmatrix} \sum_i T_{1i} f_i \\ \sum_i T_{2i} f_i \\ \sum_i T_{3i} f_i \end{pmatrix} = \begin{pmatrix} T_{11} f_1 + T_{12} f_2 + T_{13} f_3 \\ T_{21} f_1 + T_{22} f_2 + T_{23} f_3 \\ T_{31} f_1 + T_{32} f_2 + T_{33} f_3 \end{pmatrix}. \tag{13}$$

Similar to that the divergence of a tensor is written as

$$\nabla \cdot \boldsymbol{T} = \begin{pmatrix} \sum_j \frac{\partial T_{j1}}{\partial x_j} \\ \sum_j \frac{\partial T_{j2}}{\partial x_j} \\ \sum_j \frac{\partial T_{j3}}{\partial x_j} \end{pmatrix} = \begin{pmatrix} \frac{\partial T_{11}}{\partial x_1} + \frac{\partial T_{21}}{\partial x_2} + \frac{\partial T_{31}}{\partial x_3} \\ \frac{\partial T_{12}}{\partial x_1} + \frac{\partial T_{22}}{\partial x_2} + \frac{\partial T_{32}}{\partial x_3} \\ \frac{\partial T_{13}}{\partial x_1} + \frac{\partial T_{23}}{\partial x_2} + \frac{\partial T_{33}}{\partial x_3} \end{pmatrix}. \tag{14}$$

Finally the double dot product between 2 tensors results in a scalar quantity

$$\boldsymbol{T} : \boldsymbol{S} = \sum_i \sum_j T_{ij} S_{ij}. \tag{15}$$

A very important theorem concerning the accuracy of the FVM is the multidimensional Taylor Series Expansion

Theorem .1 (Taylor Series Expansion in $\mathbb{R}^d$). *Let Ω be an open subset of $\mathbb{R}^d$ and $f \in C(\Omega)^{n+1}$ with $n \geq 0$. Given $\boldsymbol{x_0} \in \Omega$ and $\boldsymbol{h} = (h_1, ..h_d)$ such that all $\boldsymbol{x'} = \boldsymbol{x_0} + s\boldsymbol{h} \in \Omega \ \forall \ 0 \leq s \leq 1$, then there exists a $\theta \in (0,1)$, so that*

$$f(\boldsymbol{x_0} + \boldsymbol{h}) = \sum_{k=0}^n \frac{1}{k!} \left(h_1 \frac{\partial}{\partial x_1} + ... + h_d \frac{\partial}{\partial x_d} \right)^k f(\boldsymbol{x_0}) \tag{16}$$

$$+ \frac{1}{(n+1)!} \left(h_1 \frac{\partial}{\partial x_1} + ... + h_d \frac{\partial}{\partial x_d} \right)^{n+1} f(\boldsymbol{x_0} + \theta \boldsymbol{h}). \tag{17}$$

Proof. See [21, §VIII] and [20, §XX.168]. □

The next one is the Gauss or Divergence Theorem which is in fact the base of the FVM and occurs in several forms and can for example be found in [20]. The exact formulation is often slightly different depending on the usage, the one used here is chosen to be suitable for the derivation of the FVM.

Theorem .2 (Gauss). *Let Ω be a compact subset of $\mathbb{R}^3$ with a piece-wise smooth boundary $\partial\Omega$ and the outer normal vector $\boldsymbol{n}$. If $\boldsymbol{F}$ is a continuously differentiable vector field on an open subset of Ω, then*

$$\int_\Omega \nabla \cdot \boldsymbol{F} \, \mathrm{d}\boldsymbol{x} = \int_{\partial\Omega} \boldsymbol{F} \cdot \boldsymbol{n} \, \mathrm{d}s_x. \tag{18}$$

Proof. A proof is for example given in [20] $\square$

Bibliography

[1] Hans Dieter Baehr and Stephan Kabelac. *Thermodynamik*, volume 15. Springer, 2012.

[2] Edin Berberović, Nils P van Hinsberg, Suad Jakirlić, Ilia V Roisman, and Cameron Tropea. Drop impact onto a liquid layer of finite thickness: Dynamics of the cavity evolution. *Physical Review E*, 79(3):036306, 2009.

[3] Jay P Boris and David L Book. Flux-corrected transport. i. shasta, a fluid transport algorithm that works. *Journal of computational physics*, 11(1):38–69, 1973.

[4] Elkin Botia Vera, Antonio Souto Iglesias, Gabriele Bulian, and L Lobovskỳ. Three sph novel benchmark test cases for free surface flows. 2010.

[5] Joseph Boussinesq. *Essai sur la théorie des eaux courantes*. Impr. nationale, 1877.

[6] Franck Boyer and Pierre Fabrie. *Mathematical tools for the study of the incompressible Navier-Stokes equations and related models*, volume 183. Springer Science & Business Media, 2012.

[7] JU Brackbill, Douglas B Kothe, and Charles Zemach. A continuum method for modeling surface tension. *Journal of computational physics*, 100(2):335–354, 1992.

[8] Roland Bulirsch and Josef Stoer. *Numerische Mathematik 2*. Springer-Verlag Berlin Heidelberg, 2005.

© Springer Fachmedien Wiesbaden GmbH, part of Springer Nature 2019
M. Jäger, *Fuel Tank Sloshing Simulation Using the Finite Volume Method*,
BestMasters, https://doi.org/10.1007/978-3-658-25228-1

[9] Ekachai Chaichanasiri and Chakrit Suvanjumrat. Simulation of three dimensional liquid-sloshing models using c++ open source code cfd software. *Kasetsart J.(Nat. Sci.)*, 46(6):978–995, 2012.

[10] R. Courant, K. Friedrichs, and H. Lewy. über die partiellen differenzengleichungen der mathematischen physik. *Mathematische Annalen*, 100(1):32–74, Dec 1928.

[11] John Crank and Phyllis Nicolson. A practical method for numerical evaluation of solutions of partial differential equations of the heat-conduction type. In *Mathematical Proceedings of the Cambridge Philosophical Society*, volume 43, pages 50–67. Cambridge University Press, 1947.

[12] S Márquez Damián. An extended mixture model for the simultaneous treatment of short and long scale interfaces. *Doktorarbeit. Universidad Nacional Del Litoral. Facultad de Ingenieria y Ciencias Hidricas*, 2013.

[13] F De la Vallee Poussin and William P Timlake. An accelerated relaxation algorithm for iterative solution of elliptic equations. *SIAM Journal on Numerical Analysis*, 5(2):340–351, 1968.

[14] L Delorme, A Colagrossi, A Souto-Iglesias, R Zamora-Rodriguez, and E Botia-Vera. A set of canonical problems in sloshing, part i: Pressure field in forced rollcomparison between experimental results and sph. *Ocean Engineering*, 36(2):168–178, 2009.

[15] Radii P Fedorenko. A relaxation method for solving elliptic difference equations. *USSR Computational Mathematics and Mathematical Physics*, 1(4):1092–1096, 1962.

[16] Joel H Ferziger and Milovan Perić. *Computational methods for fluid dynamics*, volume 3. Springer, 2002.

[17] Robert A Gingold and Joseph J Monaghan. Smoothed particle hydrodynamics: theory and application to non-spherical stars.

Monthly notices of the royal astronomical society, 181(3):375–389, 1977.

[18] Vivette Girault and Pierre-Arnaud Raviart. *Finite element methods for Navier-Stokes equations: theory and algorithms*, volume 5. Springer Science & Business Media, 2012.

[19] Ami Harten. High resolution schemes for hyperbolic conservation laws. *Journal of computational physics*, 49(3):357–393, 1983.

[20] Harro Heuser. *Lehrbuch der Analysis Teil 2*. Vieweg und Teubner, 14 edition, 2008.

[21] Harro Heuser. *Lehrbuch der Analysis Teil 1*. Vieweg und Teubner, 17 edition, 2009.

[22] Charles Hirsch. *Numerical computation of internal and external flows: The fundamentals of computational fluid dynamics*, volume 1. Butterworth-Heinemann, 2 edition, 2007.

[23] Cyril W Hirt and Billy D Nichols. Volume of fluid (vof) method for the dynamics of free boundaries. *Journal of computational physics*, 39(1):201–225, 1981.

[24] Tobias Holzmann. *Mathematic, Numerics, Derivation AND OpenFOAM®*. Holzmann CFD, 4 edition, 2017.

[25] Hrvoje Jasak. *Error Analysis and Estimation for the Finite Volume Method with Applications to Fluid Flows, 1996*. PhD thesis, Ph. D. Thesis, University of London Imperial College, 1996.

[26] WP Jones and BEi Launder. The prediction of laminarization with a two-equation model of turbulence. *International journal of heat and mass transfer*, 15(2):301–314, 1972.

[27] Dimitri Kuzmin, Matthias Möller, and Stefan Turek. Multidimensional fem-fct schemes for arbitrary time stepping. *International journal for numerical methods in fluids*, 42(3):265–295, 2003.

[28] BP Leonard. Sharp simulation of discontinuities in highly convective steady flow. 1987.

[29] BP Leonard. Universal limiter for transient interpolation modeling of the advective transport equations: the ultimate conservative difference scheme. 1988.

[30] BP Leonard, AP Lock, and MK MacVean. Extended numerical integration for genuinely multidimensional advective transport insuring conservation. *Numerical Methods in Laminar and Turbulent Flow*, 9:1–12, 1995.

[31] Yu-long LI, Ren-chuan ZHU, Guo-ping MIAO, and FAN Ju. Simulation of tank sloshing based on openfoam and coupling with ship motions in time domain. *Journal of Hydrodynamics, Ser. B*, 24(3):450–457, 2012.

[32] Florian R. Menter. Zonal two equation k-ω turbulence models for aerodynamic flows. In *24th Fluid Dynamic Conference*, 1993.

[33] Joe J Monaghan. Simulating free surface flows with sph. *Journal of computational physics*, 110(2):399–406, 1994.

[34] F Moukalled, L Mangani, M Darwish, et al. *The finite volume method in computational fluid dynamics*. Springer, 2016.

[35] William F Noh and Paul Woodward. Slic (simple line interface calculation). In *Proceedings of the fifth international conference on numerical methods in fluid dynamics June 28–July 2, 1976 Twente University, Enschede*, pages 330–340. Springer, 1976.

[36] CFD Open. Openfoam v1712 user guide. *OpenFOAM Foundation*, 2017.

[37] OpenCFD. *OpenFOAM - The Open Source CFD Toolbox - User Guide*. OpenCFD Ltd., United Kingdom, v1606+ edition, july 2016.

[38] Stanley Osher and James A Sethian. Fronts propagating with curvature-dependent speed: algorithms based on hamilton-jacobi formulations. *Journal of computational physics*, 79(1):12–49, 1988.

[39] Suhas Patankar. *Numerical heat transfer and fluid flow*. CRC press, 1980.

[40] Suhas V Patankar. A calculation procedure for two-dimensional elliptic situations. *Numerical heat transfer*, 4(4):409–425, 1981.

[41] Suhas V Patankar and D Brian Spalding. A calculation procedure for heat, mass and momentum transfer in three-dimensional parabolic flows. In *Numerical Prediction of Flow, Heat Transfer, Turbulence and Combustion*, pages 54–73. Elsevier, 1983.

[42] James Edward Pilliod Jr and Elbridge Gerry Puckett. Second-order accurate volume-of-fluid algorithms for tracking material interfaces. *Journal of Computational Physics*, 199(2):465–502, 2004.

[43] CM Rhie and W Li Chow. Numerical study of the turbulent flow past an airfoil with trailing edge separation. *AIAA journal*, 21(11):1525–1532, 1983.

[44] Henrik Rusche. *Computational fluid dynamics of dispersed two-phase flows at high phase fractions*. PhD thesis, Imperial College London (University of London), 2003.

[45] Marcus Sasson, Shuhong Chai, Genevieve Beck, Yuting Jin, and Jalal Rafieshahraki. A comparison between smoothed-particle hydrodynamics and rans volume of fluid method in modelling slamming. *Journal of Ocean Engineering and Science*, 1(2):119–128, 2016.

[46] Ruben Scardovelli and Stephane Zaleski. Interface reconstruction with least-square fit and split eulerian–lagrangian advection.

International Journal for Numerical Methods in Fluids, 41(3):251–274, 2003.

[47] A Settari and K Aziz. A generalization of the additive correction methods for the iterative solution of matrix equations. *SIAM Journal on Numerical Analysis*, 10(3):506–521, 1973.

[48] Antonio Souto-Iglesias, Elkin Botia-Vera, Gabriele Bulian, et al. Repeatability and two-dimensionality of model scale sloshing impacts. In *The Twenty-second International Offshore and Polar Engineering Conference*. International Society of Offshore and Polar Engineers, 2012.

[49] Antonio Souto-Iglesias, Elkin Botia-Vera, A Martín, and F Pérez-Arribas. A set of canonical problems in sloshing. part 0: Experimental setup and data processing. *Ocean Engineering*, 38(16):1823–1830, 2011.

[50] Joseph H Spurk and Nuri Aksel. *Strömungslehre - Einführung in die Theorie der Strömungen*, volume 6. Springer, 2006.

[51] Olaf Steinbach. *Numerical approximation methods for elliptic boundary value problems: finite and boundary elements*. Springer Science & Business Media, 2007.

[52] Olaf Steinbach. *Lösungsverfahren für lineare Gleichungssysteme: Algorithmen und Anwendungen*. Springer-Verlag, 2015.

[53] Peter K Sweby. High resolution schemes using flux limiters for hyperbolic conservation laws. *SIAM journal on numerical analysis*, 21(5):995–1011, 1984.

[54] Roger Temam. *Navier-Stokes equations: theory and numerical analysis*, volume 343. American Mathematical Soc., 2001.

[55] Grétar Tryggvason, Ruben Scardovelli, and Stéphane Zaleski. *Direct numerical simulations of gas–liquid multiphase flows*. Cambridge University Press, 2011.

[56] Henk A Van der Vorst. Bi-cgstab: A fast and smoothly converging variant of bi-cg for the solution of nonsymmetric linear systems. *SIAM Journal on scientific and Statistical Computing*, 13(2):631–644, 1992.

[57] Bram Van Leer. Towards the ultimate conservative difference scheme. ii. monotonicity and conservation combined in a second-order scheme. *Journal of computational physics*, 14(4):361–370, 1974.

[58] David C Wilcox. Reassessment of the scale-determining equation for advanced turbulence models. *AIAA journal*, 26(11):1299–1310, 1988.

[59] David C Wilcox et al. *Turbulence modeling for CFD*, volume 2. DCW industries La Canada, CA, 1998.

[60] David L Youngs. Time-dependent multi-material flow with large fluid distortion. *Numerical methods for fluid dynamics*, 1982.

[61] David L Youngs. An interface tracking method for a 3d eulerian hydrodynamics code. *Atomic Weapons Research Establishment (AWRE) Technical Report*, (44/92):35, 1984.

[62] Steven T Zalesak. Fully multidimensional flux-corrected transport algorithms for fluids. *Journal of computational physics*, 31(3):335–362, 1979.

Printed in the United States
By Bookmasters